FLEET TACTICS

theory and practice

FLEET TACTICS

theory and practice

by Captain Wayne P. Hughes, Jr.
USN, Retired

Naval Institute Press
Annapolis, Maryland

Library of Congress Cataloging-in-Publication Data

Hughes, Wayne P., 1930–
Fleet tactics.

Bibliography: p.
Includes index.
1. Naval tactics—History. I. Title.
V167.H84 1986 359.4′2′09 86-18021
ISBN 0-87021-558-2

Printed in the United States of America on acid-free paper ♾

7 6 5 4

Contents

Figures and Tables

Tables

Figures

Foreword

As a naval aviator—harboring all the biases that term connotes, forged from years of exhilaration at the reassuring sensation of a hot cat shot and the welcomed tug of the arresting gear on a dark blustery night—I am often asked, "How much longer will carriers be the centerpiece of the U.S. Navy's tactics?" It seems as if the question of the carrier as sitting duck just won't go away. A responsible answer must revolve around technology and tactics, and the pages that follow, which treat of these topics extensively, are relevant to the whole issue. The book you are about to read does a masterful job of blending technology and tactics in a historical context. Indeed, it is the most comprehensive and comprehensible of any book on tactics yet written.

Fleet Tactics appears on the scene none too soon. Phenomenal advances in technology over the last few decades have affected virtually every aspect of naval warfare in ways that have not been easy to anticipate. Technological innovations make some systems obsolescent while they are yet in the development stage, before they reach the fleet in significant numbers. This, in addition to the fact that for decades the Soviet Union has been outinvesting the United States in naval forces and defense commitment, has created in our navy a need, unrivaled by that of any time in our history, for the study and mastery of tactics.

When a naval aviator raises the gear on his flying machine he is being catapulted into another dimension, the realm of tactics, one in which he must function until his return to the carrier. Likewise, a submarine skipper passing the harbor entrance outbound sails into a realm that demands a finely honed tactical sense—in his case, as an underwater sailor conducting independent operations, intimate knowledge of his combat systems, the capabilities and limitations of his crew, and the tactical doctrine specified for his ship.

But it is not enough for an individual pilot or commander to be tactically proficient. Today, as in Nelson's time, all tactical elements must fight as a cohesive team, and tactics must integrate into the whole the best that can be brought to bear by individual units. These, in the future, will most likely include land-and space-based assets. Neither have yet achieved their full potential in the maritime environment. Space systems exist only in embryo: if and when they are developed, it will require of naval professionals tactical skills even more sophisticated than those they yield now.

The demand on today's naval tactician to be competent in the multiplatform operations of the battle force might be challenged by some who would point out that since World War II we have seen no scenario in which a major naval battle was fought or in which the use of battle force tactics was critical. Some say that the British experience in the Falklands was merely a small-scale test of tacticians' skills, and that the Arab-Israeli encounters and the U.S. bombing of Libya provide but glimpses into the broad spectrum of naval warfare. However these operations are viewed, we should guard against the insidious tendency to limit our tactical horizons to peacetime evolutions; we should instead seek to understand the ordeal of tomorrow's technological war at sea, in all its ramifications.

Regrettably, I would contend that we fall prey to the pressures of peacetime priorities, and thus over tactical excellence favor program management, systems acquisition, and—in the absorbing struggle to keep our complex weapon systems ready for action—ship maintenance. While it would be foolhardy to neglect these important areas, it is risky and not in keeping with the lessons of history to permit tactical excellence to play second fiddle to any other needs, however essential. After all, what is the naval profession about if not tactics, tactics, and more tactics? Nothing inspires more excitement in the sailor than the opportunity to exercise his knowledge of his ship and her weapon systems—to demonstrate their full combat potential.

Fleet Tactics is a treasure house of commonsensical guidelines and stimulating ideas. It is not just easy to read, but a joy as well. Captain Hughes, with his literary wand, has transformed what can be a dry topic into a fascinating treatise that will leave the reader with a desire to read it again and again. The five cornerstones—to mention only one set of principles among many in these pages—should be committed to memory and beyond; they should become instinct.

Nor has Captain Hughes failed to accentuate the most vital ingredient of tactics: leadership. At sea, as nowhere else, "men go where their leaders take them." They survive to fight yet another day and to taste the fruit of victory, or fail, together. Thus they had better learn quickly and well how to operate as a unit. War is the unexpected; and beyond a certain point, war is unimagined violence. Only the best team can cope.

Absorbed study of *Fleet Tactics* will inspire every dyed-in-the-wool sailor who has not already done so to put tactics in its proper place—first and foremost.

And now to the reader, you have a treat in store. Best wishes for the voyage ahead!

ADMIRAL THOMAS B. HAYWARD

Acknowledgments

Captain Hugh G. Nott would have coauthored this book had he lived. His contribution at the outset is beyond expression. Together we agreed that tactics was a discipline within the broader study of naval warfare; that the foundations of this discipline needed to be brought up to date; that the study of principles was inadequate, because trends, constants, and contexts also played in the equation; and that the processes of combat ought to be expressed with dynamic models. And we both shared in the failure of find a more elegant turn of phrase for our fundamental maxim of naval tactics, Attack effectively first. Hugh's presence is at the same time felt and missed on every page.

Admiral A. J. Whittle, USN (Ret.), was the first to read every word of the original draft and probably had the greatest effect on the whole of it. Frank Uhlig and Frank Snyder of the Naval War College suggested detailed improvements that were nearly as important. Among many others who sharpened the material were Vice Admiral Thomas Weschler, USN (Ret.), and Rear Admiral C. E. Armstrong on operations; Professors John Hattendorf and Thomas Hone on naval history; Lieutenant General Philip Shutler, USMC (Ret.) on littoral warfare; Doctors Joel Lawson, John Wozencraft and Michael Sovereign on command and control; and W. Robert Gerber, Gael Tarleton, Donald Daniel, and Robert Bathurst on Soviet military science.

Much credit goes to the Military Conflict Institute, which served as an early sounding board and helped me sharpen the distinctions between the processes of land and sea battle. Among its leadership, Dr. Donald Marshall, Colonel Trevor Dupuy, USA (Ret.), and Lawrence Low were unstinting of their time and advice.

Special thanks go to Captain David Clark, USN, who conceived the "Second Battle of the Nile" while together we watched on television a disastrous Army-Navy football game.

Many others provided source material, ideas, and inspiration, including Vice Admiral Joseph Metcalf III, Rear Admiral John A. Baldwin, Dr. Wilbur Payne, Dr. J. J. Martin, Dr. Milton Weiner, Professor Neville Kirk, and Captains S. D. Landersman, Lawrence Seaquist, and E. M. Baldwin. At the Naval Postgraduate School, Provost David Schrady and Dr. Alan Washburn were especially generous with their encouragement.

Ideas not communicated are seeds cast on rocks. Readers should join me in saluting those who contributed not to the substance but to the form and expression of that substance: in Monterey, Lieutenant Commander Paul Fischbeck, Ellen Saunders, Sherie Gibbons, and Ruthanne Lowe, and at the Naval Institute Press, Richard Hobbs and Connie Buchanan. My wife, Joan, gave a double measure: forebearance without untoward jealousy of The Book was one thing; serving as foil and grammarian was to walk the second mile.

Finally, Admiral Thomas Hayward not only deserves my thanks for his advice and encouragement; he also deserves the appreciation of the American navy for reemphasizing tactical competency in the fleet.

FLEET TACTICS

theory and practice

Introduction

No naval policy can be wise unless it takes into very careful account the tactics that ought to be used in war.
—Commander Bradley A. Fiske, USN, 1905

Charting Course

The last American book on the subject of naval tactics was written more than fifty years ago and published during World War II. It was a history by Rear Admiral and Mrs. S. S. Robison. For a study by an American of the art and science of tactics, one must go back almost to the turn of the century, when tactics was *the* subject debated by naval officers. In the Naval Institute Prize Essay of 1905, "American Naval Policy," then-Commander Bradley Fiske devoted twenty-three of eighty pages exclusively to tactics. It was a time when naval officers aggressively asserted that policy and strategy were not to be unfounded wishes but plans that derived from a calculated capacity for tactical success. As one French officer, frustrated by the irresolution of his government toward the kaiser's naval buildup, wrote, "Let us be economical but let us be honest. . . . It is by *objectivity*—that is, with reference to the possible opponent—that we proportion our arms; a subjective or abstract rule has no meaning. . . . If we cannot have the navy estimates of our policy, then let us have the policy of our navy estimates."*

After forty years of maritime supremacy during which Americans have become accustomed to a navy that can do everything they have asked for, it is now time to return to an understanding of how fleets win battles, so that strategy and policy will not assume the seagoing forces have more capabilities than they do.

The turn of the century was also a time when tacticians governed the direction of technology, so much so that at least one American and one Russian author incorporated technology in their definition of tactics. Issues such as the size and placement of guns, the location and thickness of armor, and the location of the conning station and signal bridge were central concerns of the tactician. Naval Institute *Proceedings* issues from the turn of the century abound in prize essays dominated by tactics. By the 1920s the U.S. Navy's General Board of Senior Naval Officers fused strategy, tactics, and the characteristics of new warships in its deliberations and all too eagerly used the Naval War College's gaming facilities as a principal tool to resolve design disputes.

The heated discourse in the world's navies during that golden age of tactical thought paid off in World War I. The great surprises were strategical, not tactical. Alfred Thayer Mahan had believed that the principles of strategy were "laid upon a rock" but that principles of tactics were more obscure because tactics depended on technology, and technology changed. He had missed the point that strategy would also be affected by new weapons. By contrast, Clausewitz thought that useful principles could be applied more frequently to tactics, and that these principles could be transformed into doctrine more readily than strategic principles.†

This book has little new to say about the principles of war. It concentrates on tactical processes, trends, constants, and contexts. Although the tactical significance of these four elements will be our subject, the reader may conjecture that in strategic studies as well

*Baudry, pp. 16–17.
†Clausewitz, pp. 147, 152–54.

the pursuit of processes, trends, constants, and contexts will bear richer fruit than a resort to principles alone.

The shelves are bare of good books on modern tactics probably because this subject is a tough one to sort out, a bit like setting out to explore the uncharted oceans and harbors of the world in the days of Magellan and Cook. Professional libraries at the Naval Academy, Naval War College, Naval Postgraduate School, and elsewhere offer scores of books promisingly described by catalog listings as tactical treatises. Few of these books live up to the billing. Most are accounts of land campaigns or battles, spanning one or more of the twenty-odd centuries from Alexander to Guderian. Most were written by historians rather than tacticians.

Or perhaps naval officers stopped writing about tactics because they were associated with maneuvers. We junior officers wondered at "tactics drills" in the 1950s: corpens and turns at five hundred yards were exciting stuff, but they were by then no more related to sea warfare than a parade ground drill was related to land warfare. When tactics were scheduled, the schedule meant maneuvers, not battle practice. These evolutions were a vestige of the day when maneuvers *were* the heart of the effective conduct of battle. As recently as 1972 John Creswell, in his preface to *British Admirals of the Eighteenth Century,* wrote, "The era of fleet tactics, the period during which the major results in sea warfare were influenced by the maneuvering in battle by rival fleets of big ships, lasted for something under two centuries."* Creswell marks the end of the era with the Battle of Jutland. But since tactics are the employment of forces in battle, then tactics exist whether or not the forces are maneuvered. Moreover, a look at the history of tactics reveals that maneuvers were conducted for the purpose of obtaining an advantage of position relative to the enemy. Although Creswell is right about maneuvers losing the central function they once served during weapon delivery, position, and especially range to the enemy, is still a vital tactical ingredient, and maneuvering is still the tool of command in establishing the position for an effective attack.

*Creswell, p. 7.

Or perhaps interest in fleet tactics waned because the world at large was looking only for battles between warships. Major components of modern fleets are land based. To control the seas and ensure their safe employment, naval forces include land-based aircraft, missiles, and sensors that search from above and beneath the surface. Land-based weapons and sensors are so important that the example I have chosen to illustrate modern tactics in chapter 10 is a battle between a sea-based force and a naval force having significant land-based components.

This book is intended to reawaken interest among the American naval officer corps in the study of tactics. First and foremost, it aims to demonstrate that there is in the art and science of naval warfare an identifiable body of tactical theory. A sound tactical debate requires contributions from historical, technological, and analytical methods and expertise.

Regarding history, students of warfare know that it is a great teacher; in peacetime all combat experience is vicarious. Still, in his study of history Mahan did us no service by placing strategic lessons over tactical lessons.

As for technology, my friend Thomas C. Hone, who is more historian and analyst than technologist, has speculated that the foremost intellectual requirement for developing good tactics is a knowledge of how weapons work. Most naval officers will agree with him. The ability to exploit the machinery of warships—the tools of their trade—has always been the hallmark of competent naval officers. After a life of observing his admirals, who could never beat the British, Napoleon wrote in frustration, "The art of land warfare is an art of genius, of inspiration. On the sea nothing is genius or inspiration; everything is positive [observable and certain] or empirical."* With a chuckle I concede this thrust. Napoleon was right, in the context of his day, but he never understood the gulf between knowing what needs to be done at sea and doing it. Since most naval officers spend more time with machinery than with tactics, I have not underscored the prominence of technological knowledge. Nevertheless, a return to tactics cannot entail a departure from technology, and so this book

*Quoted in Landersman, appendix D.

gives proper recognition (but nothing more) to technology's role of guiding and altering tactics.

And as for analytical methods, they must be both qualitative and quantitative. Here, great attention to *trends* and *constants* is a substantive acknowledgment of the role of qualitative analysis. But the more original contribution is in the realm of models. To understand tactics is to understand the dynamics of battle, that is to say, its processes and pace. One thing nearly all writers in the golden age of tactical thought placed unabashed emphasis on was geometry and numbers. Since modern officers are used to geographical and alphanumeric displays, the disappearance of that kind of quantitative reasoning is incomprehensible. In the only contemporary book on naval tactics in the literature of the free world, *A History of Naval Tactical Thought,* Admiral Giuseppe Fioravanzo makes admirable use of quantitative methods until he arrives at his last chapter, a prediction of future naval combat. There he retreats into rhetorical discourse. A perusal of current issues of the Naval Institute *Proceedings* and the *Naval War College Review* discovers hardly a numerical relationship or mathematical model and few references to results of fleet experiments to undergird or even illustrate a tactical thesis. Paradoxically, in an age of computers and operations research the essays one reads are a string of hypotheses, assertions, and exhortations.

In this book the reader will be introduced to some basic expository models. Though simple for clarity, they show that we live in an age of complex mathematical methods. When fully developed, analytical models, computer simulations, and war games are powerful tools. Modern "descriptive" models enhance our understanding. "Predictive" models help describe temporal, spatial, and organizational relationships. And even as I write there is a great upwelling of interest in the use of models and decision support systems designed to aid in the planning and execution of actual military operations.

So this book lays the three foundations of tactical study: history, technology, and analysis. They are the tools of theory. To be applied wisely by professionals in the fleet, they have to be combined with three other ingredients: operational skills, exercises and experiments, and the knowledge of a battle's immediate goals. Thus professional naval officers who must fight and win future fleet actions are my first and most important readers.

Lay Readers and the Falklands War

C. S. Lewis, the respected Oxford scholar and writer, referred to himself as a layman in theological matters. Similarly, there is a layman in naval matters. He speaks with more eloquence than the navy's blue-uniformed theologians and at his best offers wise and detached insights over the years. But if the shrewd American layman who is devoted to better national defense is going to play a constructive role, then he needs a better grounding in naval tactics. The amateur commentator is the second reader for whom this book is intended.

When the war in the Falklands was under way, some lay observers explained the action this way:

— The sinking of the cruiser *General Belgrano* showed the startlingly new vulnerability of surface ships to submarines.
— The sinking of the *Sheffield* and other British surface warships showed their striking vulnerability to air-launched missiles.
— Therefore surface warships are obsolete, especially large and costly ones.
— Fatal attacks will come without warning.
— Naval combat is becoming more lethal for the participants.
— Had nuclear weapons been used, warships would have been even easier targets.

This book is not a comprehensive guide to all the lessons of modern naval warfare. No book can be. But anyone who reads it will, I hope, reach different conclusions about the Falklands and modern naval warfare in general. In my opinion—which I intend to explain in full later—wiser conclusions about the war would follow more along these lines:

— The sinking of the *General Belgrano,* which was built before World War II, showed again that it takes modern weapons to fight modern war. The British navy outclassed the Argentine navy, notably in the category of nuclear submarines.

At sea a naval force that is even slightly inferior will usually be defeated decisively by and inflict little damage on a superior enemy. Having appreciated its inferiority, the Argentine navy was quite right to retreat into its territorial waters, effectively taking itself out of the war. Submarines were and are capable warships.

— The *Sheffield* and all three British escorts were lost in the course of successfully doing their job, which was to protect the aircraft carriers and troop ships. Because we have not seen a fleet action since 1945, we have forgotten that it is the nature of naval combat to be fast-paced, deadly, and decisive.

If a new lesson is to be inferred, it is not that warships are vulnerable to missiles, but that aircraft armed with bombs alone cannot compete with warships that carry surface-to-air missiles. The large and courageous Argentine air force was nearly destroyed in the course of sinking half a dozen British warships. Four of these ships were sunk close to the Falkland islands. In an amphibious operation such as the Royal Navy was conducting, a fleet temporarily surrenders its tactical advantage of maneuver while it guards the beachhead. Since the enemy's scouting problem is solved, the fleet is more vulnerable and has only its active defenses to depend on.

— The United States cannot allow its surface warships to become obsolete. It depends on the safe passage of merchant ships at sea and the use of the seas to protect interests overseas, with amphibious forces if necessary. These things cannot be done without surface warships. Large ships with protection such as battleships and aircraft carriers are valuable because they have staying power and can take punishment out of proportion to their size.

Before the British amphibious landing in San Carlos Sound, the mobility of the British fleet allowed it to operate safely near the Falklands. But unsupported by sea, the Argentine ground forces on the Falklands were veritably cut off from the continent, and the Argentine air force was too remote from the scene to make its considerable strength felt.

— In modern naval combat, effective scouting is the key to effective weapon delivery. Both the Argentine and British forces were crippled by inadequate scouting. The Argentine air force and the single Argentine submarine needed better reconnaissance to track, and more important, target the key British ships, which were the two aircraft carriers. The British needed better surveillance and early warning of impending attack. At least two ships were attacked while on radar picket (that is, scouting) duty at a remote distance from the main force, where they were more vulnerable.

A missile attack cannot be allowed to come suddenly and without warning. In modern sea warfare the outcome between forces with many long-range weapons will be decided by scouting and screening effectiveness before the weapons fly.

— Naval combat at sea has always been highly lethal to the participants. But the Falklands further confirmed a trend toward *fewer* casualties per ship put out of action. Modern war on land and at sea has become more destructive for machines, not men.

— Nothing about nuclear war was learned in the Falklands War. We suspect, as before, that in nuclear war fixed targets ashore will be more vulnerable than ships at sea, which are able to maneuver to evade attack.

The lessons of the Falklands War are the lessons appropriate to its nature (a maritime war) and scale (a war of limited scope). It is the kind of war the U.S. Navy is ideally suited to fight, or to prevent, but one for which neither the Argentine navy nor the British was prepared.

Other Readers

The third reader for whom this book is intended is the youngster of about thirteen years, which is the age at which the future commander at Jutland, John Jellicoe, all of four feet six inches tall, entered the Royal Navy and reported aboard the old wooden line-of-battleship *Britannia.* Genius in mathematics, music, and other fields flowers early and I want to fill the present void in the literature of tactics to stimulate interested young readers. Since there are now so many competent computer battlefield games, there ought to be at least one book available to help explain to some Nimitz or Spruance of the future why his tactics succeeded or failed.

As a boy I "learned" tactics from Lee J. Lovette, Fletcher Pratt, C. S. Forester, and anyone else whose books appeared in the 359 and 940.5 sections of the Chicago Public Library. On rainy days I would round up friends to correlate forces composed of Tootsietoy ships and to fight housewide battles (a doorway makes a great strait to focus the action). Since we couldn't afford all the ships to build a fleet—destroyers cost five cents and cruisers ten—we made huge fleets with toothpicks glued to strips of cardboard bearing the name and combat characteristics of each ship. It was a great way to spend a rainy Saturday, but I for one would have bartered my whole Tootsietoy navy for a book like this if it had existed, and all the more quickly because it is "written for the professionals." At Marshall Field's I invested twenty-one silver dollars given to me by my grandfather to buy a copy of the 1944 *Jane's Fighting Ships*. On that day Mother and King Neptune foresaw my navy destiny.

Today's youth with their captivating electronic equipment seem much more fortunate and sophisticated than we were with our Tootsietoy ships and toothpick models. But lest the electronic generation grow cocky with its advantages and expertise, a warning should be voiced. The formulation of good tactics on a home video screen is no more the measure of battlefield prowess than it is in the game rooms of the Naval War College. Tactical thought deals with hardware, things that are necessary but not sufficient. The execution of tactics on the battlefield is a matter of leadership that captures the hearts and minds of seamen. What Edison said about inventive genius—one percent inspiration and ninety-nine percent perspiration—is just as true of success on the battlefield.

There will be a fourth and uninvited reader. He is in the Soviet Academy of Science, and he is the one person I am sure will not only read but also study and dissect this book. It is the nature of Soviet military science, so systemic, quantitative, and historical in its bent, to be thorough and comprehensive in its examination of Western military literature. This reader was a sturdy, stocky, shrewd ghost watching over my shoulder as I wrote. I hope I have been suitably enigmatic in matters of current U.S. Navy doctrine. As for the candor of the rest of the book and the profit it offers him, I refer him to the assertion, on page 216, that theory does not win battles. Still, I cannot help observing that the commanders of the Royal Navy in the eighteenth century may have learned from Hoste, Morogues, and the other French theorists the very tactics by which they finally defeated the French.

Terminology*

The etymological root of the word *tactics* is the Greek *taktika,* meaning "matters pertaining to arrangement." The narrow traditional def-

*For exact definitions see appendix A.

inition of tactics is "the art or science of disposing of or maneuvering forces in relation to each other and the enemy, and of employing them in battle." In this book, tactics refer to *the handling of forces in battle.* Tactics are not studies but techniques, not an art or a science but the very action of men in battle. Thus strategists *plan,* tacticians *do.*

The definition introduced here is deliberately fuzzy.* The key words are *handling, forces,* and *battle.* Both specifically and contextually the book will clarify these terms. The phrase *fleet tactics,* synonomous with *naval tactics* in many studies, for example those by the Robisons and by Fioravanzo, deals with operations involving coordination between multiple ships and aircraft. Sometimes single-unit tactics are discussed, but only in contexts incidental to fleet tactics. In the usual framework of ground combat, fleet tactics are similar to both *combined-arms tactics* and *grand tactics,* or *operational art*, a contemporary term I rarely find it necessary to use.

Fleet tactics concern combat in which maritime issues are at stake, or in the parlance of some, with battles over *sea control* or *command of the sea.* As used here, the term *fleet tactics* excludes the special tactics of amphibious operations. It also excludes naval support of ground warfare, or projection operations, as this support is sometimes usefully called. Two points should be stressed in this regard. First, it has been relatively rare for two fleets to meet simply and specifically to gain command of the sea. In most cases one side or the other has had to deal directly and immediately with an objective on land. Thus, an amphibious operation has often worked to curtail or complicate the freedom of action of one or both battle fleets. It should come as no surprise that a fleet can lose a battle but with suitable tactics obtain its objective of protecting a beachhead or safeguarding a crucial convoy. This having been said, the scope of the present study is too limited to allow for discussion of amphibious assault; here the subject is sea battle. Second, a fleet commander must distinguish between a maritime objective and a land objective. An air strike may be used

*Some will say tactics are deployments to *win* a battle. But this may be too much to expect of tactics of inferiority. Moreover, not all naval missions are narrowly directed at battlefield victory. Sound tactics aim to achieve the *full potential* of forces.

either to suppress the threat to the seas or to support ground operations, and it is easy to confuse these aims.

Also excluded from the arena of fleet tactics are the special tactics of *guerre de course,* the French term meaning, literally, "war of the chase," and referring to raider warfare or guerrilla war at sea. Surface raiders, submarines, and aircraft operating against shipping survived by stealth and by the erosion of the enemy through a favorable exchange ratio, using solo or small-group operations. The tactics of attackers and defenders in *guerre de course* are as different as the tactics of a decisive ground battle and a guerrilla campaign.

A few old terms reappear. Prominent among them is *scouting,* by which I mean reconnaissance, surveillance, and all other means of ascertaining and reporting tactical information to a commander and his forces. The Russian word *razvedka* means the same thing for all practical purposes. *Screening,* a word of distinguished lineage, and *antiscouting,* a coined word, are used almost interchangeably. When distinctions are important, antiscouting refers to all measures to frustrate the enemy's scouting effort, *escorting* to the acts of ships or aircraft that accompany and defend valued units from enemy weapons. Escorting, as the term will be used, is one kind of *counterforce.* Thus an antisubmarine warfare (ASW) screen is primarily an escort counterforce to protect other units from submarine attacks. But by threatening enemy submarines and complicating their attempts to close and target the escorted force, the screening ships and aircraft also perform an antiscouting function, the extent and value of which are easy to underestimate.

Command and control (C^2) is the term used for the functions variously referred to as command; command, control, and communications (C^3); C^3 and intelligence (C^3I); and so on. Specifically, C^2 refers to the correlation of information received from scouting, the commander's acts of decision, and the dissemination of orders to his forces. I have used C^2 to encompass all decision support systems and many other command support systems, including all means of communicating orders, but the term excludes scouting systems and the scouting process. *C^2 countermeasures* (C^2CM) are actions taken to inhibit effective enemy C^2. *Signals warfare* is an important and useful term that crosses the line between scouting, antiscouting, C^2, and

C^2CM. *Information warfare* is a term so nebulous that it will not be used.

My definition of C^2 is neither uncommon nor universal. It has many advantages. One is to ensure that C^2 does not narrowly refer to hardware. Another is to rid the text of the comprehensive sense of C^3I, which can easily come to include all tactical actions above a one-on-one engagement. C^2 denotes not only the uses of scouting information but also the direction of or influence over the allocation of scouting resources.

One unsatisfactory term, *strategic weapons,* does not appear in these pages. The concept of strategic warfare and strategic bombing and the transfer of that concept into the realm of ICBMs are understandable but not entirely consistent. The text concentrates instead on the tactical properties of long-range nuclear weapons. When and if such weapons are used, "handling" them on a world-girdling, intercontinental battlefield will involve special tactical skills. In that ghastly battle the National Command Authority would be a tactical commander with the problem of effectively coordinating the use of air-, land-, and sea-based weapons. Although general nuclear war is not our central subject, it would appear that the paradigm of naval combat presented here (focused on attrition) lends itself much more to an understanding of the tactics of general war than the ground combat paradigm (focused on position and maneuver).

Organization

The core of this book is chapters 6 and 7, "The Great Trends" and "The Great Constants." Toward the end of the nineteenth century, when the search for principles of war had reached a climax, naval officers went about the study of tactics pragmatically, with little resort to principles. It was commonly held by students of warfare, Mahan among them, that principles guided strategic decisions but technology so altered tactics that few tactical principles would abide. The resultant dismissal of tactical theory was paradoxical: the logical consequence should have been the study of how technology brings about tactical change. There are observable trends in tactics, such as the increasing range and lethality of weapons, and there are observable constants, such as the value of concentrated firepower in space and time. An attempt to record these trends and constants is the aim of chapters 6 and 7. In addition, chapter 8 looks at new technology and its effect on new tactics.

Chapter 9 addresses the third great element affecting tactics, the great variables, that is, the *contexts* within which the tactical commander must make his decisions in wartime, such as the orders of battle on both sides and the missions of the opposing commanders. The great variables are the immediate, concrete influences on tactical decisions that no theory can settle in the abstract or in advance.

These three influences on tactics, the trends, constants, and contexts, each so different in nature, are then analyzed in chapter 10 to show how they affect modern tactics.

But as I said at the outset, a knowledge of history also contributes to wisdom in tactics. History is a fickle teacher who lets her opinionated and ill-disciplined students draw lessons as they will. The reader deserves to know how this author approached his history lessons. Chapters 2 through 5 are four vignettes that establish my slant on tactical history.

A comment is due on the evolution of these four chapters of history. My intended coauthor, Captain Hugh Nott, and I had no intention of writing history when we conceived this book. We wanted a short book that could be read at a sitting and had the impact of a squadron of dive-bombers crashing down out of the sun, carrying its message in a compelling way. But what evidence were we going to provide for the trends and constants? After considering this problem, I wrote Hugh that there had to be three chapters encapsulating three contrasting periods of tactics:

— The age of fighting sail, when tactics developed "on the job," that is, from battle to battle. Why was it that with the evidence of tactical failure before their very eyes, the eminently practical officers who led the British navy into battle had allowed themselves to be tied to a battle doctrine that patently was not working?

— The age of the big gun, when tactics were developed in the near absence of fleet actions. It was an age similar to ours, a period of tumultuous technological change in which few opportunities arose to test the new weapons and tactics in battle.

— World War II, when battleships gave way to aircraft carriers as the central weapon system of naval warfare. It was my thesis that, contrary to popular opinion, carrier aviation was brought along almost as fast as possible. I thought it would be highly instructive for today's officers who must struggle with new missile tactics to be exposed to the major tactical decisions that naval officers had to face in employing aircraft in combat.

Hugh Nott, whose appreciation of the value of historical study equaled mine, quickly agreed, and the structure of the book was almost in place.

But not quite. Not without embarrassment I admit to seeing later that there were two tactical revolutions in World War II. Along with the new weapons came the flowering of radar, sonar, and signals warfare. Thus a new chapter, "World War II: The Sensory Revolution," had to be added.

Moreover, all the time it was apparent that other things—first three, then four, and finally five things—had to be said at the outset to clear the decks for the reader. With these five cornerstones added, the book is now organized as follows:

— Chapter 1. The first cornerstone appeases the reader who would rather emphasize high courage in battle over hard thinking. The second establishes the role of tactical doctrine. The third puts the technological influences on tactics in early context. The fourth defines and circumscribes our subject, the decisive battle. And the fifth introduces the central maxim of naval combat: Attack effectively first.

— Chapters 2 through 5. The historical chapters lay a foundation for the material that follows and give the reader who knows naval history or thinks he does a chance to compare his own interpretations of history with those here. In particular, these chapters contrast the methods of maneuver, of control, and above all of concentrating force practiced by (1) ships of the line, (2) battleships, and (3) aircraft carriers. Three different models of battle help the reader visualize the dynamics of naval actions.

— Chapters 6, 7, 8, and 9. These cover the trends, constants, and contexts (or variables) of tactics. Now we are prepared to apply the historical lessons to develop an understanding of the key processes of naval action: firepower, maneuver, scouting, and C^2.

— Chapter 10. This discusses modern tactics, along with a little model of combat in the missile age. It emphasizes two points: while visualizing a concentration of force against force is always the best way to grasp the essence of tactics, effective scouting is the key to delivery of firepower; and C^2 directs the tactical commander's decisions regarding force, counterforce, scouting, and antiscouting measures employed against an enemy who is taking similar measures to strike first.

— Chapter 11. Some concluding comments are offered here.

This book, though it cannot be read in one sitting, is meant to be short, essential, and enticing, so that naval officers—even, and especially, those busy keeping their ships and aircraft operating—will read it. But the ulterior motive is to generate an overwhelming compulsion to put tactical reading, writing, and debate back where they have always belonged: first in the minds of every line officer.

1

Five Cornerstones

Prologue

We are with Vice Admiral François Paul Brueys d'Aiguilliers in his flagship *Orient*, anchored at Aboukir Bay, Egypt. Less than ten years ago, when our revolutionaries decapitated the leadership of the king's navy, Brueys was a mere lieutenant. Now he is a protégé of General Bonaparte. Our admiral is good at organization, having achieved the prodigious feat of escorting four hundred troop transports with thirty-six thousand French infantry, cavalry, and artillery to Egypt. Still, if there was any serious danger today, 1 August 1798, we would be ill at ease.

Brueys, having achieved our Napoleon's confidence, is now having trouble living up to that trust. One feels the tension of his inexperience: he has refused to enter Alexandria, just ten miles away, for fear of grounding a ship, and for weeks now, while Napoleon has swept aside all opposition and become master of Egypt, he has vacillated between fighting his thirteen ships of the line under way or here at anchor, in the western crescent of the flat Aboukir coastline. There will be a fight in the end, because this *diabolique* Nelson is running pell mell all over the Mediterranean looking for us. Brueys has (we think) made a decision of sorts: he will fight at anchor. *Bien.* But why not get on with the plans to fortify Aboukir Island to our north? Half a dozen six-pounders in place are nothing. When one fights pinched against a coast one builds up the shore batteries. In our position shore batteries to strengthen the van are no luxury, because our admiral (sound though we know his plan to be, *mon ami*) has put his strongest ships in the rear of his column. Or rather, we should say, at the southern end of this line of ships riding at anchor. Something about the British aim to double on the rear, an argument we do not grasp because we are immobile. Or has it something to do with the prevailing wind? At this hottest season of the year, an onshore breeze to push the ships of Nelson toward the rearmost, southern end of the column?

And there are the other things agreed on but never accomplished: using stern anchors to keep the starboard batteries all in line to seaward; running cables between ships to prevent the enemy from breaking through our line and raking us fore and aft in the passing; tightening up the anchorage, for we are 150 meters apart, not much closer than a column under way, safe for swinging with the tide and convenient for boating but too far apart for ships that will be fought at anchor. Ganteaume, our chief of staff, worries about food and water and we are grateful, but we have so many men ashore foraging—three thousand, we've heard—that there's no one left with the energy (the heat, it is so oppressive!) to make these preparations. In the end we may yet get under way: Brueys himself says nothing is certain in war—we are *toujours flexible*—and how much longer before we must up anchor and seek a place with more food? If we did we could practice ship handling under way. How did we get in this mess? Du Chayla, who knows more about seamanship than any of us, says we should fight under way. But the captains resist: the men are too raw and untrained. So what do we do? Atrophy at anchor for a month and forget more and more of our training in everything but stealing Egyptian camel meat.

Stop grousing? *Oui,* it is the heat. Nelson's fleet will be inferior. The *Orient,* beneath our feet, has 120 guns. She is a monster. Not one English ship by itself can stand up to us for fifteen minutes. Just

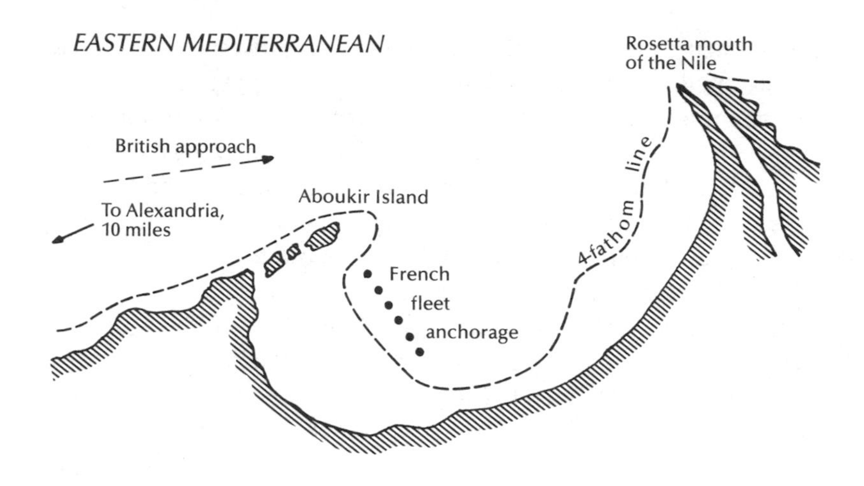

Fig. 1-1. Aboukir Bay

ahead is Admiral du Chayla's *Franklin* (who was this Franklin, *citoyen?*) of eighty guns, and astern is *le Tonnant,* of eighty. Nelson has seventy-fours and fewer of them.

Attention! The *Heureux* has signaled a dozen sail! It is the English, coasting just east of Alexandria. We see the sails now, above the flats to the northwest. *Eh bien.* We will find out now about this battle. We are galvanized. Up goes the signal for the general recall of our foragers ashore; most will be back by morning. Clear for action. One side only? It is because of all those sailors ashore. Clear to starboard, you idiot, larboard is to landward. We may not be able to sail with the shopkeepers, but we will outfight them. When? Tomorrow, at first light, a day-long fight it will be and hard.

Brueys paces the quarterdeck. Six bells in the afternoon watch and they have not cleared Aboukir Island. Why don't they haul off? The afternoon breeze is brisk and a sailor wants sea room before dark. Signal in the air from their *Vanguard.* The madman Nelson has seen our impregnable column and will stand clear now. An ounce of sense and he would blockade. That would be a pretty pickle for Brueys. In time we'd have to sail out and fight and we sail like lubbers. But Nelson, he has no patience. Tomorrow we fight. There will be little sleep tonight.

Mais ecoutez! The *Zealous* in the lead is wearing round into the bay! A man in the chains—they are sounding. Nelson is coming in *now,* standing before the afternoon sea breeze. We look to Brueys. *"Impossible!"* he cries. He is a fighter, but he has not *le sang-froid.* How many men are back from the abominable desert? Where are the cables? The stern anchors? Now the pitiful batteries on Aboukir Island fire. They are a great nothing.

So we will have a night action. No one fights at night. It will be mass confusion for ships under way. We are at anchor. Night action is crazy, but it is crazier for the English madman. Nelson's line wears around the point, starboard tack now, sailing almost west. The van! Nelson is aiming for our van, not the rear. The weak van. And he will not wait; the wind is with him. Why don't the English fall into disarray? They can sail, these sea demons of Nelson. Closed up two hundred meters between ships, yellow stripes outlining the gunports, two and three tiers of them, thirty-five muzzles from each of their black hulls. Closer, closer they come, and the hairs stand on the back of the neck and the hot sea breeze of the late afternoon has a cold chill in it. Shouts drift across three miles of bay water. The shopkeepers are cheering. Another signal from the *Vanguard.* The *Goliath* passes the *Zealous* to windward, her mainsails gathered up and then her topgallants, battle canvas only for the last rush, which she now leads.

The *Guerrier* and the *Conquerant* let go their starboard broadsides! *C'est magnifique!* A raking position, but at half a mile? We glance at Brueys, his hands gripping the bulwark, chest high. He is shaking his head. "*Non, non.* Too soon, too soon!" he cries out.

Closer, closer sail the *Goliath* and the *Zealous* and the ten other English ships of the line. In our van activity will cool the passion of fear. No place to hide in ships: where the captain fights, you fight. Back here we must stand and tremble a little, for the pall of death is staring at us. How beautiful, how relentless, how awful. Why do they not waver? Our line outguns them. Ah! Now I see why. Our line is out of the action, all but the first half-dozen ships. Nelson is doing the impossible; he masses on the van. It is his stupid luck, nothing but luck, to have this sea breeze catch us within only hours of our being ready.

The *Goliath* is almost upon the *Guerrier*. The Englishman must wear his ship to larboard now, to return fire with her first, her starboard, broadside. Nelson's line will pass down our line front to rear exchanging broadsides. In the dark what else can he do? One broadside each, twelve ships. Heavy, but the *Orient* with a sixty-gun broadside will pulverize them. Stupid Nelson. He plays our game.

Mais qu'est-ce que c'est que cela? We do not look at the *Goliath*'s starboard battery; it is her larboard battery that stares at us! Can this be so? We look at Brueys. He is chalk-white. He grips the rail, his eyes aflame, aghast. This terrible fighter has fear? "*Merde,*" he murmurs, in the voice of our doom. The *Goliath* passes in front of the *Guerrier*, inshore of her, raking at twenty yards. The *Goliath* fires with her larboard battery! The British will double on the van and destroy us ship by ship. We feel the screams, the blood, a sickening desperation in the van, and now we see death creep down the line, ship after ship . . .

* * *

Before he dies Brueys will suffer further tactical humiliation. Nelson's ships do not remain under way. Each anchors by the stern alongside the enemy, two of the Royal Navy for one of the French. Successive ships from the rear will pass on the disengaged side of the British ships ahead and anchor alongside the next French ship astern to the south along the French column, four altogether finding their way to the landward side where the French are without preparation and defenseless. As the nothern French ships are smashed, the British will make their way down the line, always with the firepower of two or more against one.

The French fought at the Nile with the passion that Napoleon inspired. The French flagship *Orient* handled the English *Bellerophen* and *Majestic* so brutally that the *Bellerophen* cut her anchor cable to drift clear, the *Majestic* was dismasted, and her captain was killed. But meanwhile the *Alexander* slipped cannily between the *Orient* and the *Tonnant*—which at anchor in Aboukir Bay were too much separated to give fire support or prevent penetration of their line. The *Alexander* then devastated the *Orient* from landward, almost unopposed. The *Orient* caught fire, and the flames spread to the magazines. At ten that night the French flagship blew up. No one who saw it ever forgot the stunning, soul-searing terror of it.

The southern French ships, five of them, were frozen at anchor, out of the action. To this day we do not know whether this was because they were awaiting a signal sent from Brueys to join the action, which was never seen in the smoke and darkness, or whether it was because the rear admiral, Pierre de Villeneuve, had no time to cut cables, reach painfully to windward, and join the action. We only know what he wrote later. In his report he asked how his ships, "moored by two large anchors, a small one and four cables [could] have weighed and tacked to get within range of the fighting before the ships engaged had been disabled ten times over." Villeneuve, far from being destroyed professionally by this debacle, would command the French and Spanish fleet at Trafalgar. But it is said that his spirit was broken at the "Horror of the Nile" and that he and his fleet, in facing Horatio Nelson a second time, were morally beaten before the first broadside was fired.

Of paramount importance, Nelson had done what Brueys had not: he had spent his two months at sea not merely drilling his force but discussing and planning the battle. His captains all knew his plans for concentrating firepower by doubling. Exactly how would depend on circumstances: whether the French were at anchor or at sea, and if at sea, whether they were to windward or leeward. He was to write before the Battle of Trafalgar that "something must be left to chance," but nothing was left undone that could be foreseen. His fleet would attack at once. At the Nile it was fortuitous for him that Brueys had three or four thousand men ashore. Nelson did not know this; he could not know *what* profit he would extract by a readiness to go in on arrival, but all of his forces were inculcated with the belief that in battle time is precious.

At the Nile the payoff was tremendous. Besides being shorthanded and therefore needing desperately to fight the starboard batteries only, many of the French ships were at anchor only by the bow. We are not certain whether it was Nelson or Captain Foley in the *Goliath* who first saw that a ship anchored only at the bows must have space to swing and therefore must leave sailing room ahead to pass across to its landward side. Either way, it was Nelson's emphasis on concentration that inspired the move. His original plan for fighting ships at anchor envisioned one ship on the bow and one on the quarter of each unit of the engaged enemy. It was the temper of his captains to implement his intent by doubling in the more conventional manner of ships under way by placing vessels on either side of each ship of the French van.

We may also believe that Nelson attacked at once because he had an afternoon sea breeze and he understood that the wind would die toward evening. He had to accept a night action in exchange for a speedy attack and a favorable wind. That Nelson contemplated night action is plain from the provision for a series of horizontal lanterns in the rigging that would distinguish friend from foe, and from his plan to anchor from the stern instead of remaining under way. The final, masterful touch was his plan, well drilled in the mind of every captain, to reverse the order of ships, the first two British ships to engage the northernmost French ship at anchor, the next two taking the second, and so on, in shrewd recognition of the fact that coming down on an enemy at anchor was entirely different from closing when he was under sail and in motion. Some of Nelson's captains thought his plan could not be executed. They said if one ship had two ships concentrating on it, the two would mask each other's fire. It was a risk Nelson accepted and drilled to overcome. And with his own towering knowledge of seamanship he could talk down all reluctance.

As figure 1-2 shows, Nelson's plan was distorted beyond superficial recognition in the execution. So it is with battle. But Nelson's captains never wavered from his intent. A good battle plan is a simple plan. It will leave room for subtle and complex considerations, for variations, and for both error and initiative at the moment of execution.

Our imaginary Frenchman who described the battle's onset was not exaggerating his armada's capacity and will to fight. At the Nile the French fought with the fervor of men unaccustomed to losing. They had no seamanship, but they could pour out broadsides. Against first-class fighting men, which the French of 1798 were, tactics established the circumstances of victory, while will power and skill translated the potential into reality. The battle raged—it is the appropriate word—all night and into the dawn. The British had nearly

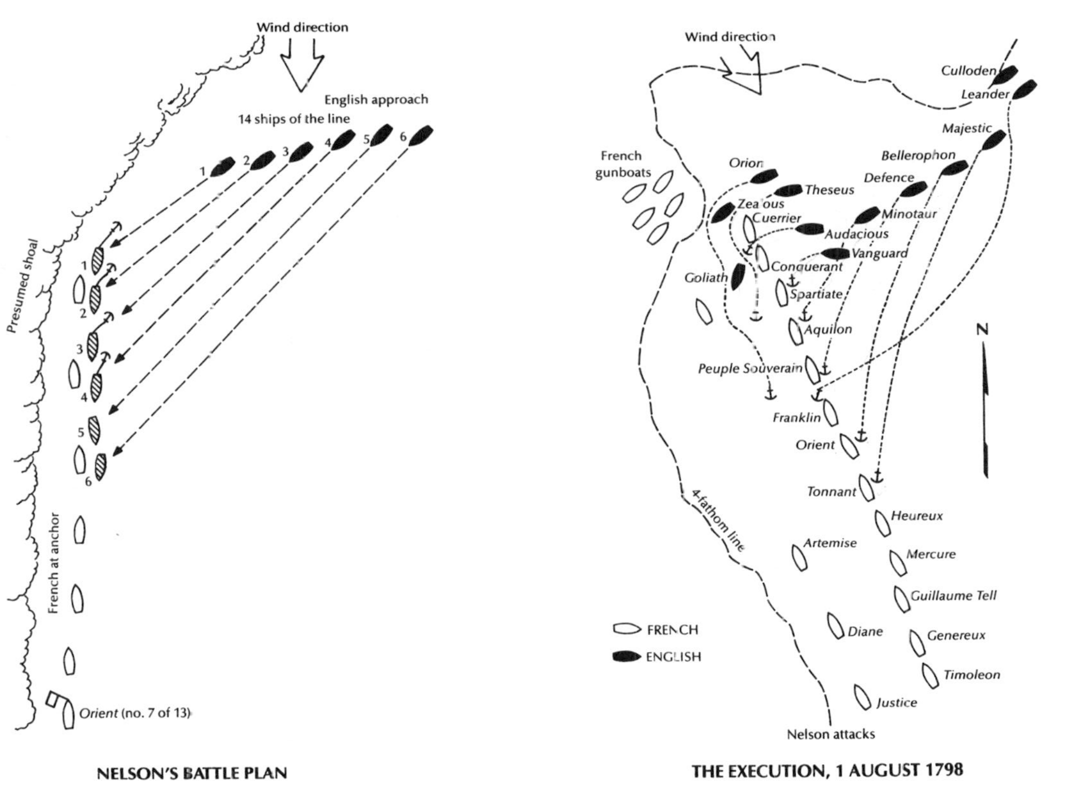

Fig. 1-2. Royal Navy Maneuvers at the Battle of the Nile

one thousand casualties. The French had over three thousand, and they lost another three thousand prisoners.

The Five Cornerstones

The Battle of the Nile illustrates five crucial points about maritime warfare that have affected naval tactics in general and in particular:

— Leadership, morale, training, physical and mental conditioning, will power, and endurance are the most important elements in warfare. One cannot win without quantitative and qualitative sufficiency of material and sound tactics to bring one's weapons down upon the enemy, but—and this is especially important in regard to an enemy with competitive means—*men matter most.* At the Nile Nelson established tactical ascendancy at the outset, but his force also had to win a very hard battle. Sound tactics, extensive training, and careful planning will win a great victory when they are accompanied by an unwavering determination to inflict—and accept—losses.

— Doctrine is the companion and instrument of good leadership. It is the basis of training and all that that implies: cohesion, reliability in battle, and mutual understanding and support. Further, doctrine is the springboard and benchmark of all tactical improvement. Formally, doctrine is standard battle methodology. But doctrine is not dogma. No one is esteemed more than Nelson for his readiness to unfetter the Royal Navy from the doctrinaire rigidity of its permanent fighting instructions. But Nelson always had a plan of action, a comprehensive one. He always transmitted it to his captains and practiced it so that they were of one mind about what was wanted. On a fundamental level, there can be no better definition of battle doctrine than a comprehensive and practiced plan of action. Sound doctrine will establish unity amidst chaos. So although in one sense the Battle of the Nile did not go as planned at all (the reader will look at figure 1-2 in vain for an execution pattern that resembles the prebattle plan), in a deeper sense it is the epitome of a sound plan executed flawlessly in spirit. The nineteenth-century Prussian army leader Helmuth von Moltke said,

"No plan survives contact with the enemy." Nelson understood as well as any man that *doctrine is the glue of good tactics.*

— Tactical and technological developments are so intertwined as to be inseparable. That is why Mahan rejected (rather too readily) constants of tactics while he promoted principles of strategy. The Battle of the Nile occurred near the end of the age of fighting sail. Nelson had little opportunity to adopt tactics to new material, as Napoleon did with mobile artillery and the great Panzer captains did with tanks. Thus Nelson's achievement is even more remarkable: he adopted his tactics to a weapon system that in its essentials was centuries old, and with insight that has rarely if ever been equaled at sea. We may believe that his tactical mastery was achieved by a lifetime spent under way. Clausewitz thought that good strategy could come from the inspired novice but that effective tactics were the work of a lifetime. *To know tactics, you must know weapons.*

— While it is proper to think of the destruction of the enemy's fleet as the fleet's foremost objective, beyond that immediate objective is always some higher goal. *The seat of purpose is on the land.* That is another reason the Battle of the Nile is so appropriate an example. It was fought inshore, as so many decisive battles have been, almost in the harbor, with one side at anchor and the other partly so, and with shore batteries playing a part that could and should have been greater. The battle devastated Napoleon's ambitions and would have destroyed a lesser man. The battle that nearly obliterated a fleet also, and more significantly, destroyed an army's sea line of communications.

— The tactical maxim of all naval battles is *Attack effectively first.* This means that the first objective in battle is to bring the enemy under concentrated firepower while forestalling his response. We will pass lightly over this, the greatest imperative of sea warfare, because it will be discussed at length later. Suffice it to say that, with the readiness of his force and his instant decision to bore in at once and exploit the unreadiness of the French, Nelson was adhering to this maxim. But as we shall see, it embodies much more than the element of surprise, mental and material readiness for immediate action, or even the spirit of the offensive.

Men Matter Most

Warfare is deadly conflict. Tactics, being the devices of battle, are conceived and executed at the physical and metaphorical center of this violence. Tactics are more visceral in the consummation than policy, strategy, operational art, or logistics. A debate has continued for two centuries over the extent to which the practice of warfare is a science or an art. Both sides in the debate understate the role in warfare of what may be called the mystique or charisma of leadership. To read the earliest literature on warfare, one would think that this was the major force behind victory in battle. Has so much changed? Science and art are activities of the mind; neither reflects what, insofar as the battle itself is concerned, transcends both in importance: the matter of will and endurance, and among combat leaders the ability to inculcate these qualities in their forces.

This book is not about the inspirational qualities of leadership; it treats tactics as a matter of the mind. But beneath the veneer of reason lie passion and mortal danger. Nothing about battle may be understood without grasping its violence. Poetic imagination is not a quality with which we military men are heavily endowed. If we were, we might all be mad. It was said of C. S. Forester's greatest fictional hero, Horatio Hornblower, that had he really lived, he would have been put ashore with ulcers. Still, I hope that my vignette of the Battle of the Nile has evoked a picture of the human element of warfare, of emotion elevated to a feverish pitch, and of the way tactical plans and battlefield decisions are influenced by an environment of controlled violence and directed chaos.

I have no way of judging Napoleon's statement that "the morale is to the material as three is to one." But whatever may be true on the ground, in naval warfare the ratio is probably narrower because in ships at sea the men go where the leaders go. In planning naval tactics it is best to assume, with occasional exceptions, that there will be equal wit, valor, and perseverance on both sides. This is an important assumption. Sun Tzu, who wrote *The Art of War* three centuries before the birth of Christ, said the greatest of commanders win by outwitting the enemy, by outmaneuvering him, and even by leaving open for him a line of retreat so that he will be encouraged to break and run from the battlefield. Liddell Hart, the prolific champion of maneuver warfare after the bloody stalemate on the Western Front in World War I, believed that the best tactics involve ingenuity and avoid head-to-head battles of attrition. But his famous philosophy of the indirect approach applies to battles on land; it has a hollow ring to a commander at sea. Naval battles are hard fought and destructive. It is, however, possible to overstate the significance of courage and high morale in combat, which is dependent on machines. This the Frenchman Ambroise Baudry did in his treatise on naval tactics at the turn of the century. He was doubtless influenced by his compatriots in the army, who had a zealous commitment to élan until they found in World War I that on the ground an offensive mindset can be overdeveloped. At sea the greater danger is of a misplaced faith in the sufficiency of tactical cleverness. If the enemy's morale or intellect is inferior, so much the better; but to base tactical success on outwitting, outmaneuvering, or outfighting a first-class enemy is the height of folly. The tactician does what he can to place his forces in advantageous circumstances for battle, but not without admitting the possibility of the tables being turned. Even on land, officers like Ulysses S. Grant and the British general Douglas Haig who have been accused of military butchery had, on the contrary, recognized that the enemy—the Confederates in the first instance and the Germans in the second—were determined soldiers led by men of great tactical skill. Such forces are not defeated without bloody battle. Superior tactics may tip the balance, but the wit and ingenuity required for such tactics become overshadowed by sheer grit in the latter stages of a long war.

At sea the predominance of attrition over maneuver is a theme so basic that it runs throughout this book. Forces at sea are not broken by encirclement; they are broken by destruction. Over the years naval strategists have been careful about committing their forces to battle at sea because of its awesome destructiveness. Compared with land warfare, major sea battles have been few and far between. Partly this is because, the estimation of material superiority being relatively easier to gauge at sea than ashore, naval strategists in the inferior navy have tended to avoid battle until the jugular vein was threatened. A superior navy with a modest force advantage has often been able to contain and neutralize a strong enemy and carry out many strategic objectives—up to a point—without fighting. Considering the death and destruction wrought by naval warfare, it may be that the very decisiveness of battle at sea, which leads to the avoidance of it, is a virtue for which the civilized world can be grateful.

Doctrine Is the Glue of Tactics

The second cornerstone of naval tactics is doctrine. Doctrine is the commander's way of controlling his forces in writing, before military action. Doctrine enunciates policies and procedures that govern action. In its broad sense, doctrine is what is taught as "right behavior"—"rules upon which we act spontaneously and without orders for the accomplishment of the mission," wrote Admiral Harry E. Yarnell.* In its stringent sense, doctrine is mandatory behavior; it must be obeyed. Either way, high levels of command want coherence of policy for control, while tactical levels want procedures for cooperative effort. These are merely matters of emphasis. Two points about doctrine must be remembered: it is vital, and it must not become dogmatic.

Fleet doctrine may be thought of as the commander's comprehensive battle plan—his standing operation order. (Nelson's plan of action is an exemplar.) Every echelon of combat command has its battle plan. One of the anomalies of modern American organization is that the chief of naval operations prescribes a comprehensive doctrine, his naval warfare publications, but he is not in the operational chain of command.

Doctrine may also be thought of as every action that contributes to unity of purpose. Doctrine isn't what is written in the books; it is what warriors believe in and act on. Clausewitz called it "a sort of *manual* for action."† Doctrine is greater than tactics in that it encompasses command structure and communciation. It is less than tactics in that it can establish no more than procedures that enable and enhance the execution of tactical choice on the battlefield.

*Quoted in Robison, p. 827.
†Clausewitz, p. 141.

In the execution of good doctrine there is always tension between conformity and initiative. Draw any good naval leader, senior or junior, into a conversation on his experience in naval operations and it will quickly come out that the tactical action imposed by his seniors was to his mind too rigid. He will tell you how he maneuvered more cleverly and fired his weapons more effectively than doctrine prescribed. In the next breath he will tell you how when he was in command his units moved together like clockwork. He will swear to you that all his captains knew exactly what each teammate would do as instinctively as a basketball player knows from body language which way his teammate will cut.* It will never occur to the speaker that there is the slightest inconsistency in his account. To a man, strong military leaders want freedom for initiative from their seniors and reliability from their juniors. Doctrine in the hands of able commanders will, at its most sublime, allow the achievement of both these things. There is a measure of entropy in all doctrine. With too little entropy there is order and understanding but no initiative. With too much entropy there is creativity and change but no order.

The sublime rarely being possible, if one must err it is better to do so on the side of too much rather than too little doctrine. Too little doctrine indicates laziness, indecisiveness, or uncertainty. The clearest evidence of doctrinal deficiency is too much communication—reams of orders and directives which in the planning stage are little more than generalities and exhortations, and which defer too much to the moment of decision. Good doctrine reduces the number of command decisions in the heat of battle, for even a cool head will be gripped by passion and, very quickly, emotional and physical exhaustion.

American naval officers today are—it is difficult to pick the right word—wary of doctrine. This is nothing new. The excessiveness, rightness, or meagerness of doctrine is so much a part of tactical discourse that the Robisons' *History of Naval Tactics* might well have

*The immediate source of this analogy is Lieutenant General John Cushman, USA (Ret.), one of the ablest of our writers on C². But many have used sports analogies. The first man I know of to relate sports to tactics (not to esprit, which goes back to the Iron Duke and his playing fields of Eton) was W. S. Sims, who likened drilling for war to plays in a football game.

been entitled *The History of Naval Doctrine;* their stress is very much on systems or orders, commands, and signals. They recount the active debate that started when the U.S. Fleet was big enough to make controlling it in battle an important matter. Then as now, doctrine tended to mean what the user wanted it to mean. Rear Admiral Robison attempts to referee and interpret the various viewpoints rather than declaim his own, but at one point he expresses his attitude in this way: ". . . It [the term *doctrine*] gradually fell into disuse [about 1915]. It is probable that 'doctrine' would not have been offered as a naval term if 'Fighting Instructions' had been in existence, separate from signal books."* In 1981, when he was Commander Second Fleet, Admiral James A. Lyons promulgated fighting instructions for his fleet. His successor, Vice Admiral Joseph Metcalf, was deeply interested in developing them further. During the 1920s and 1930s the fleet tactical publications expanded into combat doctrine. Little analysis has been published of their effectiveness in World War II. Most commentary is, as one might expect of observations after the fact, critical of their flaws, but it is evident that the publications were believed in and acted on, the sine qua non of doctrine. They provided the foundation of training, gave the fleet standard operating procedures, and were the point of departure for tactical developments.† Today we use the Allied Tactical Publications of NATO, the Naval Warfare Publications, and a plethora of transitory documents. Whether their content is now mature is a question for officers in the fleet to answer in light of the following indicators of utility.

Doctrine is the basis for training and for the measurement of the achievement of training standards. Doctrine is, on the one hand, the tactical commander's assurance that when a new combat unit reports to a fighting force, it comes already equipped with certain combat skills needed by the commander. Doctrine is, on the other hand, assurance to the captain of the new unit that his ship will fit quickly into the new fighting force and that he and his men will not have to adapt to a bewildering set of new signals and procedures on the eve

*Robison, p. 827.

†One well-researched evaluation of fleet tactical publications and their effectiveness in World War II is McKearney, "The Solomons Campaign."

of battle. Doctrine provides continuity of operations when captains are transferred or killed.

Doctrine provides the commander's staff with a basis for comparison of their force with the enemy's. Force evaluation requires the association of weapons with tactics. Any evaluation is, of course, a good deal more than the comparison of orders of battle, but if the comparision—in Soviet parlance, the correlation of forces—cannot be made, no battle outcomes can be estimated, and without them no strategy can be drafted, and without that defense policy is built on sand.*

Tactical doctrine is the standard operating procedure that the creative commander adapts to the exigencies of battle. It is the procedure from which a ship's captain knowingly departs to exploit an opportunity, fully confident that his fellows will act in a predictable way; indeed, in the best of worlds with the best of captains, even the departures seem predictable in the circumstances. Paradoxically, doctrine generates initiative: a trained subordinate can see from it not only what will be done but what will not, and know as Nelson did at Cape St. Vincent how to save the battle.

Doctrine is the procedure to which the tactician may propose a change, and the standard the tactical analyst uses to evaluate the new against the old. It describes the way present weapons are employed so that their effectiveness may be compared with that of a new weapon. It is the basis for understanding the new tactics that accompany new weapons, that is, for understanding the reward of prospectively greater effectiveness, the penalty of training in new techniques, and the burden of undergoing the transition from old weapon doctrine to new.

In sum, doctrine must be whole and firm but not dogmatic. It must leave room for men of freewheeling genius, for such will be the aces of the next war. But it must never surrender control, because control is the prerequisite of concerted action. Although in the most difficult circumstances, the creator of doctrine may find that control alone

*Some talk as if the national policy determines national defense policy, which in turn determines strategy and tactics. This scheme is a rational way to approach force procurements for a *future* strategy, but present policy, like wartime aims, depends on present means.

will not be enough to win, control comes first and will count for more than inspiration in the midst of battle.

To Know Tactics, Know Technology

Universally recognized are these two facts: technology advances keep weapons in a state of change, and tactics must mate with the capabilities of contemporary weapons. The American navy in particular has been fascinated with hardware, esteems technical competence, and is prone to solve its tactical deficiencies with engineering improvements. Indeed, there are officers in peacetime who regard the official statement of a requirement for a new piece of hardware as the end of their responsibility in correcting a current operational deficiency. This is a trap. Former Atlantic Fleet commander Admiral Isaac Kidd, Jr., was always a champion of the need to be prepared to fight with what you have. And no wonder: his father died fighting in the *Arizona* at Pearl Harbor.

The tactician stays ready by knowing his weapon systems. Technical facility, like good leadership and sound doctrine, is the third cornerstone of this book. We bow to the great god technology and honor him as a jealous deity who will wreak vengeance on all apostates.

Our ablest naval officers were tacticians who knew their technology: William S. Sims with his continous-aim fire; Bradley Fiske with his host of patents, including one for aerial torpedo-release gear before aircraft were even capable of lifting a torpedo payload; and William A. Moffett and other early aviators who foresaw the day when naval aircraft would be potent ship killers and who helped develop bigger engines, navigating equipment, carrier arresting gear—all the machinery to fulfill their visions.

The great historian of the Civil War, Douglas Southall Freeman, condensed the ten commandments of warfare into three: "Know your stuff; be a man; and look after your men."* His first commandment dealt in large part with competency in battle tactics and field equip-

*Freeman's speech of 11 May 1949 is reprinted in the *Naval War College Review*, March–April 1979, pp. 3–10.

ment. But what is true in ground combat, where machines serve men, is magnified at sea, where men serve machines.

The Seat of Purpose Is on the Land

In the classical manner, this book centers attention on fleet actions. This is right and proper. Command of the sea, which today includes the air above the surface and the water beneath it, is still the requisite of the effective employment of sea power. We have come to the end of a thirty-year era during which the U.S. Navy could take sea control largely for granted and concentrate on how to exercise that control with operations that project naval influence ashore. The purpose of this book is to examine and highlight the tactics of fleet-on-fleet action, but especially because of its emphasis on sea control, it is important to say at the outset that sea battles are not fought for their own sake.

For one thing, the study of maritime history reveals that fleet battle is rare; the landing of force, the support of operations ashore, and the protection of shipping at sea are the most common employments of navies. It is worth pointing out that conducting an amphibious operation, clearing a minefield, or escorting a convoy were never easy propositions.

For another thing, and this is even more to the point, the great decisive sea battles between fleets have always been connected with events on land, usually in an immediate, direct, and obvious way. These connections are matters of strategy and therefore outside the purview of this book. I will, however, link tactics with mission and strategic purposes later to show that peacetime tactical development derives from anticipated wartime roles. The only certainty about our navy's wartime role is the uncertainty of predicting in peacetime what site, enemy, and mission will be involved. This makes the problem of developing tactics acute. The modern means of achieving direct naval influence are as follows:

— Direct attack with SLBMs, nuclear or conventional cruise missiles, aerial bombs, or naval gunfire.
— The augmentation of land combat operations with air support, naval gunfire, and riverine warfare.

— Isolating an enemy with a naval blockade, attacks on shipping, or offensive mining operations.
— The initiation of land operations with amphibious assault.
— The protection of military reinforcements and resupply.
— The protection of economic sea lines of communication. The protection of shipping is as important in peacetime as in wartime. If a major war occurs, these trade routes will be distorted beyond recognition, most likely never to resume their prewar structure.
— Peacekeeping presence or "deterrence." Presence is only as effective as the apparent willingness to exercise force. Presence should be appreciated not only for its military and political value but also for its economic value in fostering favorable trade and investment.

It has long been axiomatic that naval influence cannot be exercised before "sufficient" control of the sea is secured. The classic way to secure control is to defeat the enemy's means to contest that control, his main force or "fleet." Today the potential is greater for tactical interaction between land and sea elements, including aircraft, missiles, and long-range sensors. A naval force-on-force action need no longer be ship on ship. The greater range of land-based aircraft and missiles over the last fifty years has required a reinterpretation of "naval" forces. Clear thinking is required to determine whether the purpose of an intended battle, which may be over continent or ocean or both, is continental or oceanic—that is, whether it is projecting force from the sea (implying a measure of control over the land) or controlling the sea (implying the threat of attacks from the land).

Attack Effectively First

The first four cornerstones of maritime warfare (as distinct from naval tactics) have been introduced in part to obviate the need for a running apology.

Yes, men will dominate battle.

Yes, doctrine, though not the same as tactics, must be sound to facilitate tactics.

Yes, weapons influence tactics deeply.

Yes, the decisive sea battle is a means and not an end in itself.

The fifth cornerstone is different: it is the tactical theme running throughout this book. The great naval maxim of tactics, Attack effectively first, should be thought of as more than the principle of the offensive; it should be considered the very essence of tactical action for success in naval combat.

Fifty years ago the Robisons ended their long history of tactics by observing that the most important tactical maxim is Attack. It is a strange conclusion; most everyone else regards the fundamental tactical goal to be the concentration of force. We have seen the disaster for the French of a blind faith in élan and the offensive in land battle. We know also of Clausewitz's great respect for defensive positions. Let us agree that there is wisdom underlying the Robison conclusion, note that we are dealing with fleet tactics and nothing more, acknowledge that selecting the time and place of battle bestows an obvious advantage on the attacker, but for the moment defer consideration of the Robison enjoinder, Attack.

In 1914 Frederick W. Lanchester introduced his celebrated equations to show the consequences of concentration of force in the modern era.* The narrow physical consequences, expressed quantitatively, yielded a square law of effectiveness. Lanchester compared these results with a linear law (refined later by other authors) governing two forces constrained in their ability to concentrate firepower.

The physical effect of concentration was phenomenal. Lanchester said that "ancient conditions" (linear law conditions) of limited weapon range and mobility pitted individual against individual in what was equivalent to a series of duels. If the fighting value of individual combatants was the same on both sides, 1,000 fighting men meeting 1,000 enemies would result in a draw; but if 1,000 men were concentrated against 750 in a battle of annihilation, 250 of the larger force would be left when the smaller force was eliminated. Superior concentration would lead to victory, but barring psychological effects, the larger force would suffer casualties equal to those of the smaller. Under "modern conditions," however, a new advantage favors the side with larger numbers. When both sides can aim their fire, there is an expanding, cumulative advantage for the larger force. At any

*Lanchester was a successful British automotive engineer with an intellectual curiosity that swept over aerodynamics, economics, fiscal and industrial policies, the theory of relativity, and military science.

instant, the rate of casualties imposed is proportionate to the numbers of forces remaining, and the ratio of forces continuously increases in favor of the stronger initial force. In simplest form, if the attrition rate of force A is proportionate to the remaining strength of B and vice versa, we have attrition equations

$$\frac{dA}{dt} = -B$$

and

$$\frac{dB}{dt} = -A$$

omitting for clarity the killing rate parameters for both sides, so that in these equations the fighting effectiveness of individuals on both sides is the same. The solution to these coupled equations is

$$A^2_o - A^2_t = B^2_o - B^2_t$$

where A_0 and B_0 are the inital force strengths and A_t and B_t are the strengths later, at time t. If A has 1,000 fighters each of whom can fire at any enemy, and B has 750 fighters with the same capability, when the battle is fought to the annihilation of B, about 660 of the A side will survive, rather than 250.

Lanchester then asked, What would happen if the fighting quality of one side's forces was superior to that of the other side's? He showed that for aimed fire, numbers of fighting units are more valuable than fighting quality. A commander is better off with twice as many units of force than with units with twice the rate of effective firepower.*

Uninformed writers have carried the application of Lanchester's equation further than he intended. Critics have with reason taken some of these extensions as too literal or too ephemeral. There is irony in this, because the square law treats only the dynamic physical effects of homogeneous firepower. Lanchester wholly set aside the

*Lanchester's ingenuity was in recognizing the simpler form that could be used to express the mathematical effects of concentration previously quantified by Bradley Fiske and others. Fiske also saw that collective fighting strength was more important than individual fighting value. A contemporary of Lanchester's, the Russian M. Osipov published similar equations in 1915 and explored their applicability to ground combat.

possibility of the enhanced effectiveness of synergism.* Lanchester did not dabble in this, nor did he deal with the comfortable psychological advantage that numerical superiority usually bestows.

Adaptations of Lanchester's equations have been applied commonly in ground combat analysis but infrequently in naval combat. This is because land battle has been fought by larger numbers of forces, which are very difficult to analyze except in some simplified and aggregate form. I want to suggest that, with suitable modification to allow for ships that can take hits and continue to fight, his exposition of cumulative advantage is more relevant to sea than land, because the opportunity to concentrate firepower in the manner he specified is more likely to occur at sea. Ground combat is characterized by position, movement, and ultimate considerations of territory. Rather than attack effectively first, perhaps the best maxim for ground commanders would be Nathan Bedford Forrest's statement, Make it a rule to get there first with the most men. The subtle distinction is the same as that in chess between pawn-takes-pawn and pawn-to-king-four.

The potential to *effect* this concentration is greater at sea than on land. At sea there is no high ground, no river barrier, no concealment in forests that requires what is often used as a rule of thumb on land, a 3:1 preponderance of force to attack a prepared position. As others have said, battle at sea and in the open desert have much in common. Sun, wind, and sea state all affect naval tactics, but not to the extent that terrain affects ground combat. It is because of this that attacking has not carried the penalty of sea that is imposed ashore. Over the course of history, the central problem of naval tactics has been to attack effectively, that is to say, to bring the firepower of the whole force into battle simultaneously.

A second and subordinate objective of naval tactics has been to try to concentrate one's whole force on a portion of the enemy's in order to defeat him in detail. For the inferior force, this was a necessity. For either side, superior or inferior, massing in the face of

*That is, of using combinations of weapons in such a way that the effectiveness of the whole is greater than the sum of the parts, or mathematically, the set [A + B] is greater than set [A] + set [B].

an enemy who could *see* the opposition was so tricky that the superior commander usually declined to try, and the inferior commander vastly preferred (mission aside) to decline battle or cripple his enemy so that he could withdraw. But it was evident that a concentrated first strike by one force against a part of the enemy's was at once an opportunity to exploit and a hazard to avoid.

In the days of fighting sail and the rifled gun, tacticians sought ways to achieve decisive first attack by maneuver and technologists sought ways to achieve it by weapon range and ship maneuverability. In World War II the importance of attacking first swelled. In daylight hours aircraft were more valued than guns because they outranged them by an order of magnitude. The importance of timing reached a new peak. Since the range of action of carrier-based aircraft on both sides was comparable in 1942, who launched the first effective attack became a matter of superior reconnaissance and intelligence—better scouting. Land-based assets for scouting and attack also played a key role. In the great Pacific carrier battles, what was foremost in the carrier fleet commander's thinking was the paramount advantage of getting off the first concerted attack.

To attack effectively (by means of superior concentration) and to do so first (with longer-range weapons, an advantage of maneuver, or shrewd timing based on good scouting) have been the warp and woof of all naval tactics. Everything else—movement, cover and deception, plans, and C^3—has been aimed at achieving such an attack. We may now understand why the Robisons chose to distill their wisdom into the one word *attack,* even as we reject it and consider our own maxim with great care.

Attack effectively first, like all general truths, is not very helpful in specific circumstances. As a maxim of naval tactics, it is most useful not for what it says but for what it excludes and leaves unsaid. All fleet operations based on defensive tactics (but not all defensive forces) are conceptually deficient. A successful defensive naval strategy entails a concentration of force and a successful tactical attack.* Effective fusion of reconnaissance, surveillance, and intelligence informa-

*Excluded as inappropriate here is the strategy of a maritime *guerre de course,* which is discussed in chapter 9.

tion is so important that it must receive the same emphasis as the delivery of firepower. Contrarily, obstructing the enemy's scouting by cover, deception, confusion, or distraction merits enormous attention, for successful scouting and screening are relative to each other and a matter of timeliness.

Nothing about naval combat is understood if its two-sided nature is not grasped. Each side is simultaneously stalking the other. Weapon range is relative to the enemy's weapon range. The weapon range that matters is the productive range—that is, the range at which a telling number of weapons may be expected to hit their targets. The weapon range that matters, for a battle force, is the range at which enough weapons can be aimed to hit with great effectiveness.

Our maxim cannot be reduced to the principle of the offensive. Defensive *forces* and *operations* are very much a part of operational plans. Mere ambition to attack does not preclude the enemy's success. At times the enemy tactical commander will have the means to attack at longer range. If so, one's aim is to see that the enemy cannot attack so well that a reply will fail.

Left to itself, the spirit of attacking first will fritter away weapons piecemeal; left to itself, the spirit of the offensive will marshal great magazines of weapons to serve as a funeral pyre. The man who believes the quicker rapier will always win will lose to the man with a strong shield and battle ax.

The man with the quickest aim and the farthest vision will teach the wrong tactics to a man less well endowed: the enemy will have his own Drake, Suffren, or Tanaka, and tactical theory must teach steady leadership first because genius will take care of itself.

2

Tactical Development in Action: The Age of Fighting Sail, 1650–1805

Introduction

The two and a half centuries of naval warfare that have come to be called the age of fighting sail were a time when sailors had for their tactical laboratory the battlefield itself. In this chapter we will observe a strange phenomenon of the age: the ossification of British tactics for nearly a century. And we will examine how it came about that hard-headed naval fighters became handcuffed by what was more than doctrine, the entire system of C^2, the permanent fighting instructions. We will see the frustration of eighteenth-century British admirals created by a rigid interlocking system of tactics, doctrine, and communication.

In eighteenth-century battles, the ship of the line was the means of concentrating firepower and the line ahead was the most practical means of unleashing that firepower cooperatively. As many naval writers explain, it was natural for sailing ships to fire abeam, so a line-ahead formation was logical. Since the admiral's communications by flags were hard to see, control was simplified when the commander located his flagship in the center of a column. With a column he could maneuver a large number of ships, not swiftly, but with a minimum of confusion and with a minimum of communication. The direction and force of the wind and the bearing and distance of the enemy determined his maneuvers. He deployed in full view of the enemy, and sometimes breakfast was served before battle. His objective was to bring all of his force, "well ordered, well knit, and simultaneously,"* with no unengaged reserve, against the enemy. If through training and seamanship the admiral could place a tightly spaced column alongside a raggedly disposed enemy line, he would succeed in augmenting his collective firepower wherever enemy ships were spread thinly, wherever they overlapped and masked each other's fire, or wherever some were out of line and out of effective range.

The Ship of the Line: The Means of Massing Firepower

It took great seamanship to bring the firepower of even two ships simultaneously against a single enemy. Fully effective gun range was well inside three hundred yards, at point-blank range. Maximum significant gun range was eight or nine-hundred yards; beyond this the probability of hitting was remote and the hull penetration of roundshot was poor. The firing arcs of broadsides were limited to about twenty-five degrees fore and aft of the beam. Training a gun was a slow and awkward process, so that, by and large, guns were trained by maneuvering the ships. Thus it was rare for two consecutive ships in column to enjoy the advantage of directing fully effective fire at a single enemy simultaneously. Figure 2-1 illustrates how, at the outer limits of effective range—about five hundred yards—it was just possible for two ships closely spaced in column to direct broadsides against an enemy if his column was not closed up.

Another way to increase the density of firepower was to stack cannon vertically. Hence the logic of two-deckers and three-deckers. This was the most effective method of massing force at sea. By the late seventeenth century three-deckers were common, and naval leaders knew their great practical value. In 1697 the highly respected Frenchman Paul Hoste would write that the size of vessels contributed

*The felicitous words of Creswell, p. 178.

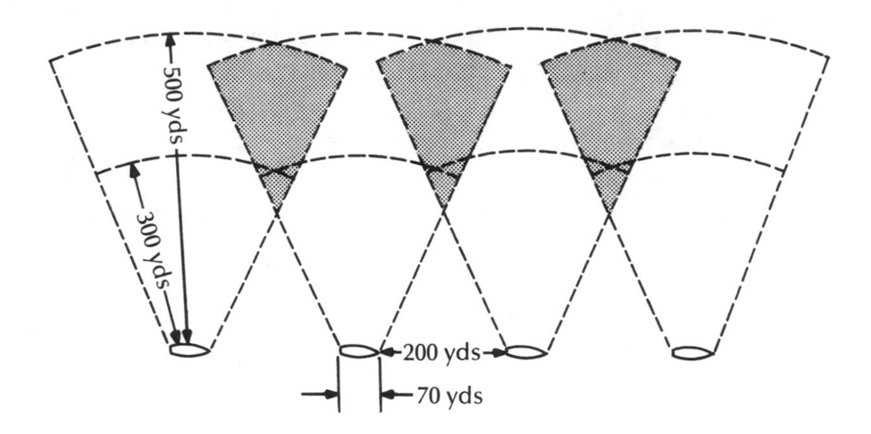

Fig. 2-1. Effective and Point-Blank Range of Ships in Column

1. Ships are at 200-yd intervals, extremely tight.

2. At point-blank range, 300 yds, ships in column are just able to present a solid front of broadsides to ships trying to break through the column.

3. At a reasonably effective range of destructiveness, about 500 yds, adjacent ships have some power of concentration (shaded areas), but not much.

more than numbers to the strength of the fleet for two reasons: larger ships had more and heavier guns, and a fleet of large ships (as opposed to a more closely spaced column of smaller ships) brought to bear more and larger guns in the same length of column.* Single-deck ships diluted the density of firepower too much to sail in the line, and even two-deckers were, by the late eighteenth century, being put in the line with reluctance.

We may presume that the two-decker was observed in action to have much more than a 2:1 advantage over a single-decker (in due course, a frigate). It was the firepower advantage that mattered most: Hoste spoke of the defensive strength of a larger ship's timbers, but he understood that this was a lesser advantage. Two ships fighting

*Robison, p. 220.

each other at sea were much the same as two stockades on wheels, with wooden walls protecting—somewhat—the guns and gunners. Barring a lucky shot in a magazine, ships were not often sunk by gunfire (though they sometimes sank later from hull damage). Ships were defeated by what today we would refer to as a firepower kill—by knocking out guns and gunners and by crushing morale and the will to fight. As many ships were captured as sunk. Since British seamen fought better and French ships had excellent sailing qualities, there were many rehabilitated French ships in the Royal Navy.

If the two stockades were armed with the same number and quality of men and guns, the rates of fire, accuracies of fire, and shot penetration would decide the winner. On the other hand, if a two-decked seventy-four (with a thirty-six-gun broadside) met a frigate of thirty-eight guns (an eighteen-gun broadside), the "square-law effect" would result and the frigate would suffer great cumulative damage relative to the seventy-four. No amount of prowess could be expected to save the frigate in a stand-up fight. With simple continuous fire on both sides and no hull strength advantage for either, the two-decker would lose only five guns while destroying the frigate's entire broadside of eighteen, according to the computation. Let us suppose that sometime—not later than the destruction of his entire broadside—the captain of the frigate surrendered. A force of thirty-eight guns and three-hundred seamen would have been lost at a cost to the two-decker of five guns or less and similarly light casualties and structural damage.

If a two-decked seventy-four were matched against a three-decker rated at one hundred guns, then, using the same model and assumptions as before, we can see that the two-decker would be forced to capitulate before the three-decker had lost twenty guns. In the North Sea and the English Channel this cumulative advantage of a modest edge in initial firepower was seen in action. One suspects that many overcritical courts-martial of two-decker captains transpired, as well as scarcely warranted knightings of three-decker captains.

The Column: The Means of Control

It is proper therefore to think of the column itself primarily as a means of controlling the force inherent in the admiral's ships and only secondarily as a means of effecting concentration of firepower. This was exactly the way the admirals of the latter half of the seventeenth century thought of it. In fact, two-decked fighting ships were already numerous in the first half of the century, when the concept of the commanded column was instituted. Admirals knew, however, that mutual support resulted from a close interval between ships, and the interval specified in their fighting instructions could be incredibly (and unobtainably) short, as little as three ship lengths between ships.* It took great seamanship to keep even a short line of a dozen ships at close interval, especially when ships had to avoid overtaking and masking the fire of vessels ahead. Keeping station was difficult at best, but it was made easier in one respect. Ships under "fighting sails"—the middle sails (topsails) alone—advanced at stately speeds of about three hundred feet a minute, bare steerageway. Crews were needed at the guns. Cohesion and control in action as well as firepower were valued more highly than speed.

A second reason for tight spacing was to prevent an enemy from breaking through the column and raking ships on either side from stem to stern at point-blank range with entire and unanswerable broadsides. An additional possibility was that after some ships had broken through the column, usually at the rear, which was to windward, that portion of the column would be enveloped by ships on both sides, victim of the process known as doubling (see figure 2-2).

Finally, a closely spaced column was desirable in case the enemy, by design or accident, came against part of it with his whole force. Unengaged ships, usually downwind in the van, would have to beat upwind as quickly as possible to get into action, and for them, distance meant time.

Thus, whether he was dealing with single ships or a commanded column, for three centuries the tactician's problem was to concentrate firepower at sea in an era when effective gun range was very short, less than five hundred yards. The C^2 solution was to form a firing line, to keep the closest possible interval without risk of overlap, and

*Distance was specified in cables (240 yards long). In the Anglo-Dutch wars, when ships were fifty yards long, a distance of half a cable was specified in at least one set of instructions. Later a prescribed interval of two cables was common (Robison, pp. 153, 361).

The aim of doubling was to achieve one of these positions:

Both fleets were moving, at about the same speed when well handled, so it was difficult to attain the positions without splitting a fleet temporarily and risking the enemy's closing on one half or the other and countercon-centrating. Nelson successfully doubled on the French at the Battle of the Nile because they were at anchor and other circumstances were fortuitous:

British under way

French at anchor

The plan to double on the rear, depicted below, illustrates the difficulties. In the first sketch victim fleet *B* is sailing close hauled on the starboard tack. Doubling fleet *A* is squared off, ship for ship, just outside of effective range to windward.

In the next sketch the ships of fleet *A* have backed sail until the half in the van are abreast of the rear half of *B*'s line. As *A*'s van closes to engage, the rear of *A*'s fleet falls off the wind to sail in echelon under the sterns of *B*'s passive column.

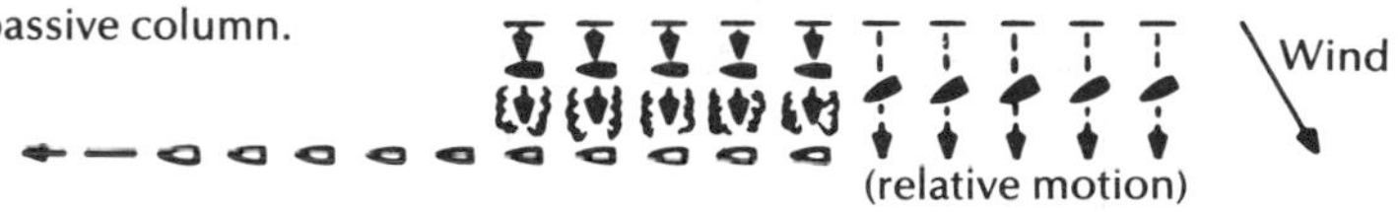

Wind

(relative motion)

With *A*'s van now closely engaged with *B*'s rear, the rear of *A*'s line now to leeward, hauls up close to the wind. *B*'s engaged rear is presumed to be moving slowly, allowing *A*'s leeward ships to move up into doubling position. *B*'s unengaged van must wear around, and if it takes too long it won't gain the battle.

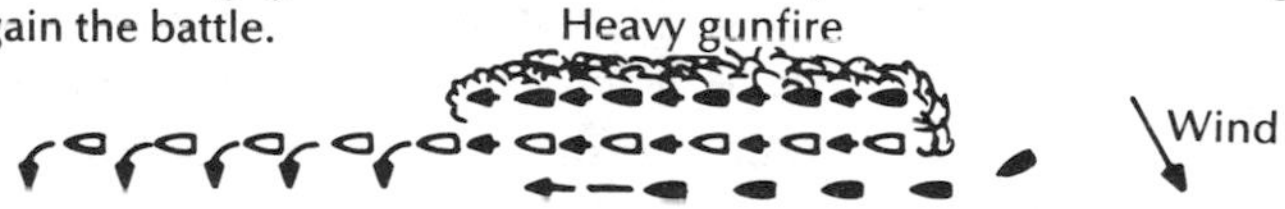

Heavy gunfire

Wind

It doesn't take an 18th-century seaman to appreciate that even a lethargic fleet *B* could itself back its sails and spoil the plan, or fall off downwind to disengage from *A*'s windward ships and frustrate those attempting to get to leeward. Doubling on the van would be less propitious still, because *B*'s unengaged ships in the rear would automatically sail forward and enter the fray, even in the absence of orders to do so.

Fig. 2-2. Doubling

to stack firepower vertically. With this in mind, let us briefly review the evolution of tactics under sail in order to understand the reluctance of British naval leaders to tamper with the great tactical strength of the line.

The Creative Phase

We will take as our starting point the incorporation of the commanded column into the English fleet admiral's fighting instructions. The three Anglo-Dutch wars of the latter half of the seventeenth century were a ferocious competition over trade. During the first war no less than six major sea battles were fought in a mere two years (1652–53). One reason there were so many battles was strategic. We will return to this later. Another reason was technological; at the outset of the wars fighting ships doubled as commercial carriers and could be built or repaired quickly. But tactical factors are the main explanation for so many battles; because cruising formations were clusters, the actions ended in tactical indecision. Ships were polyglots of size, shape, and sailing qualities, and captains, who doubled as tradesmen, were undisciplined in formation sailing. At the outset there were too many ships and there was too little control. Flight of the enemy from the field was the measure of victory, not the number of ships sunk or captured. Evidence suggests that when the enemy was sighted an effort was made to form loose columns, each under a flag officer, but once the battle was joined it became a back-alley brawl. After the indecisive third battle near Portland, Oliver Cromwell sent three of his generals, Robert Blake, Richard Deane, and George Monck, to sea. Accustomed to order and discipline, they jointly issued new instructions that made columns the basic formation, with the understanding that ships would form on each flagship separately, in several lines.

There is no evidence that concentration of firepower was the initial motivation, except insofar as control lent itself to concentration. Bearing in mind that the ships began the war as armed merchantmen, with captains patriotic and piratical but not naval, and that the flag officers were landsmen new to the sea, the columns were probably conceived as the simplest way to control the fleet. Then, as now, it was folly to devise a battle plan that the force had not been trained to execute, and keeping interval in even a short column of ships under sail was no mean accomplishment. But to the extent that the formation was maintained, the tactic succeeded. It achieved four things:

— Control of each column by its admiral, who used very few signals, each keyed to an article of the commander's own fighting instructions and appropriate to the level of skill and training.
— Avoidance of the danger, frequently encountered in the smoke and confusion of a melee, of firing into friendly ships. The confidence that this danger was removed would have been as important as the fact. One would not have wanted to take time for gunsmoke to clear to be sure he had an enemy in his sights.
— Elimination of shirkers. Article 7 of the fighting instructions warned potential recalcitrants that the severest punishment awaited captains whose ships failed to form column. We may believe this was not a parade-ground injunction.
— A set of commanded columns, at least three, which offered promise of getting all ships into action and under control in the shortest time. There was to be no reserve. A ship at a distance was not viewed as a reinforcement to be thrown in at a critical time or place; a ship out of range was wasted firepower.

The Fighting Instructions Mature

The admirals understood quite early in the age of sail that the close column also concentrated fleet firepower. As the Royal Navy sailed and drilled and fought it was welded into a professional force of fighting seamen. Firepower was increased by increasing the rate of fire and by keeping station at close interval with superb seamanship. By the sixth and last battle of the First Anglo-Dutch War, the Dutch were forced to form column too. Because the Dutch came to fight, the Battle of Scheveningen was bloody and decisive and the war ended.

In the Second and Third Anglo-Dutch wars the single column became the standard formation for the fleet, but maneuvering could be adapted to circumstance. Both sides being aggressive and skillful, the English admiral could be content with a single column and devote his fleet's skills to gaining the weather gauge, upwind, since the line to windward could come down and fight or stand off and decline battle. While the English were putting more and more faith in the single line, the potential of the windward fleet to mass or double on the enemy was not lost on the combatants.

But for the English the single column, or *fleet* line ahead, would soon be too much of a good thing: a formation designed for beginners, it became a rigid doctrine before the new breed of naval professionals could develop and exploit the full C^2 potential of the fighting instructions. It was indisputable that a tightly knit column maneuvered by its admiral was the way to achieve local concentration. But when the whole fleet—as many as one hundred ships—was formed and maneuvered as a unit, the result was a thin ribbon of firepower, miles long but less than five hundred yards deep on a side. It took an hour for the rear ship to arrive at the position of the foremost. Worse yet, depending on the wind, the van might not be able to succor the rear at all.

English Tactics Ossified

At the opening of the eighteenth century, fleets were so large that it was nearly impossible to bring a single long line, that thin ribbon of destruction, simultaneously against a reluctant enemy, van to van, center to center, and rear to rear. The eighteenth-century French were such a reluctant enemy. Taking their motivation to have been the avoidance of a decisive action, we can say the French succeeded, by concentrating on accuracy instead of volume of fire, arming their ships with more long-range guns, positioning themselves to leeward, and firing high into the rigging to cripple English ships, forcing them to drop out and weakening the line. But the main reason the French succeeded in avoiding decisive battle was that by 1740 the English admirals' own fighting instructions had been supplanted by what came to be called the Admiralty's permanent fighting instructions, which the commander in chief was permitted to supplement, but with which he thought he must, upon pain of death, comply without substantial departure.

As Professor John Hattendorf of the Naval War College pointed out to me, John Creswell offers recent evidence that until 1799 the fighting instructions were not, in fact, Admiralty instructions, contrary to the common belief attributable to Mahan and Sir Julian Corbett. Creswell contends that there was nothing basically wrong with Royal Navy tactics. To the extent that this is true, it reinforces my argument for coherent, practical doctrine. But whether the "Sailing and Fighting Instructions for His Majesty's Fleet" were issued, endorsed, or merely noted by the Admiralty hardly seems to have mattered: the concept of the line, fought preferably to windward, was drummed into the minds of the fleet commanders in chief. And fighting with the coterminous line, the Royal Navy suffered nearly a century of inconclusive battles and frustration.

The permanent fighting instructions were not only doctrine but dogma. As well as being the approved means of controlling the fleet, they prescribed all signals controlling maneuver. The fleet could hardly conceive, much less comply with, creative tactics: they simply were not provided for in the instructions. What could not be practiced could not be executed in battle. Concentration on the enemy rear, breaking the line, or doubling against a part of the enemy came to be opportunities the admiral could not exploit, indeed, could probably not see at all, because they had been eliminated from his vocabulary, which was embedded in fighting instructions. As the eighteenth century wore on, this C^2 albatross stifled not only tactics but tactical imagination as well.

Naval Tactics Restored

We venerate Horatio Nelson not for breaking free from the rigid formalism of the permanent fighting instructions. George Rodney (1782), Richard Howe (1794), and Adam Duncan (1797) did that, and John Jervis shares the glory for appreciating the significance of Nelson's tactical genius at the Battle of Cape St. Vincent (1797). Nelson is our peerless tactician because he knew how to exploit the Royal Navy's new-found freedom. I am speaking now of accomplishment. Corbett gives credit to Rear Admiral Richard Kempenfelt for drawing up new orders and signals that permitted maneuvering flexibility as early as 1780. Robison (p. 346) implies that the problem of too much formalism was widely understood, and that the means to solve it were already fermenting. The French were, as usual, well ahead of the British in tactical theory and their writings were thought to have inspired better British understanding. For sound practice and aggressiveness, rare in Frenchmen of his time, we can still admire Admiral Pierre Suffren's tactics two centuries after his five actions in the Indian Ocean. The person who implements a new tactic is rarely the inventor. We honor the former because he has the vision to pluck a kernel of wisdom from a barn floor of ideas and the courage and skill to make it grow.

It has been said that Nelson stands out in our memory because the Battle of Trafalgar, fought on 21 October 1805, was the last big fleet action for more than a century. But his fame does not rest on an accident of history: there were no more fleet actions because he did what no one else had been able to do. He eliminated the enemy fleet, thereby ending the need for further fleet actions and setting the stage for a hundred years of British naval dominance. We know Trafalgar was not a lucky stroke; Nelson, with his experience at the Nile and at Copenhagen, understood concentration and timing. We know this also from the instructions he issued to his captains before the battle. They are often quoted, but no one has analyzed them better than the Frenchman Ambroise Baudry.* Baudry directly addresses the time and motion elements and Nelson's contingency plans. He points out that the French admiral Villeneuve knew almost everything Nelson would do and yet did not know how to stop him. Villeneuve could not conceive a counterinitiative.

Nelson need get little credit for winning, even though the numerical odds were his twenty-seven against Villeneuve's thirty-three. That he would win as much of the battle as he could was a foregone conclusion. Nelson's accomplishment was to destroy the French and Spanish fleets. That had been the British strategic objective for ten years. He succeeded as no other admiral had for one hundred years.

His problem was to close and grip the enemy, whom his captains would destroy if they got within that deadly three-hundred-yard range. He chose the right tactics to defeat a particular enemy and thereby achieve his country's particular strategic objective. He had a plan, he communicated it, and as events proved it was demonstrably adaptable even during execution. His ships were trained for his plans—to unleash a high rate of close-range fire—and his plan fit their training.

*Baudry, pp. 218–36.

His captains knew what to do because he talked to them at every opportunity and because the plan in its essence could not be misconstrued. Lastly, his plan could be executed from beginning to end almost without signal. The order of sailing will be the order of battle—this notion, a pertinent watchword for a modern fleet, was a stunning innovation in 1805.

One can learn the wrong lesson from Trafalgar. Nelson's tactics against, say, Marten Tromp or Michiel de Ruyter or Suffren would have been disastrous. Every one of his ships in those light winds of October had to run a gauntlet of three or four unanswered broadsides, a guarantee of defeat had his opponent been first-rate. His ships would have been half crippled before they fired their first effective shot. Nelson would have known this. It is the shallow interpretation that we caution against, the one from which it might be concluded that Nelson won simply because he attacked boldly. A lesser commander who has not measured his enemy carefully will destroy his fleet by parroting Nelson's charge into the enemy's fire. We do not know how Nelson would have fought the Dutch in 1688 or the Germans in 1916, but we may trust that his tactics would have been right for the strategic objective, the weapons, and the forces assembled.

The Influence of Strategy

It is necessary to comment briefly on national policy and strategy in order to understand the tactics of the British, the Dutch, and especially the French. Tactics influence and are influenced by strategy. The tactical commander must never forget his aim in battle. That aim will often extend beyond the destruction of the enemy fleet.

In the three Anglo-Dutch wars both nations' wartime objectives were centered on the oceans, and as a result many battles were fought. The strategic issues could have been summed up in Monck's characteristically blunt reply when he was asked the cause of the second war: "The Dutch have too much trade. . . and the English are resolved to take it from them."* There was no question of invasion; trade was the issue, and it was contended at sea. For the Dutch, commerce via the English Channel was survival. With it she pros-

*Quoted in Michael Lewis, p. 89.

pered; without it she would wither into nothing. Neither side could decline battle and achieve its purpose in the war. It either built up its navy and fought, or made peace and lost its objective. The wars had limited objectives, so that the winner could indeed anticipate a net financial gain. The loser could anticipate financial and national ruin. Hence the motivation to fight and fight to win. Both sides knew that the decisive battle would settle the war until a new fleet was constructed, at which time there would be a new battle.

It suited France's strategic purposes to decline decisive battle. British advantage lay in winning a decisive battle and exercising command of the sea. But the Royal Navy had learned its tactical lesson too well against the aggressive Dutch. Like tactics failed against the reluctant French. For them, the ocean was a flank to be held while the decision was fought out on land. Whenever they believed their war aim would be determined on land, one of two naval strategies was usually adopted. They would maintain a substantial fleet to divert the British navy (not always successfully) and to look for opportunities. Or they would conduct a *guerre de course*, hoping for moderate gain at little cost.

Whether or not France's naval strategy was wise is not an issue for this discussion. But concerning her tactics, we may observe the following:

— They were, on the whole, more successful in implementing French strategy than British tactics were, before Nelson, in implementing British strategy.

— When great benefits would have accrued to the French by aggressive, decisive battle, the possibility was virtually out of reach. Long-practiced tactical habits ran deep.

— When the Royal Navy learned how to close and fight a decisive battle, the French, who wished not to fight, were devastated tactically and therefore severely hampered strategically.

Summary

Because decisive weapon range was very short, tactics in the age of fighting sail were always dominated by ship-on-ship combat. Until nearly the end of the era, effective concentration was improved primarily by building up the weight of broadside per ship, or rate of fire, or accuracy of fire. The British built up rate of fire; the French, accuracy of fire.

The commanded line was invented in the 1650s to achieve coordination (rapid maneuvers in formation with minimal communications) and cooperation (movement at close quarters without shooting at one's own ships and without shirkers). Very quickly the commanders saw that multiple columns also afforded tactical concentration of firepower: ships could be brought into action together and in mutual support, so that the opening firepower of the whole force could be massed.

The single fleet line ahead had the appeal of simplicity, but in the early eighteenth century British naval doctrine was frozen into dogma. With insignificant exception, the commander was required to maintain a single rigid line, inhibiting the possibility of tactical surprise or achieving fleet concentration by doubling on part of the enemy. Worst of all, the fighting instructions limited the ability of the commander to close and defeat a reluctant enemy, even when enemy firepower was distinctly inferior or strategic circumstances demanded that he take tactical risks to force the action. From the middle of the eighteenth century, the last fifty years of fighting sail were, for the admirals of the Royal Navy, dominated by an effort to escape their doctrinal straitjacket and bring about a decisive engagement against an enemy who was inferior in close action.

When the fleet commanders finally freed themselves of the stricture of the single line ahead, they had still to rediscover a means of fleet concentration that prevented enemy escape. Nelson did so, and in one final stroke, Trafalgar, he eliminated the enemy at sea. His successful tactics involved:

— Operating in mutual support by column and by ship in column until the battle was joined and the inevitable melee occurred.
— A simple and unequivocal plan which still allowed flexibility of execution.
— Minimal signals because, from experience, Nelson's captains knew his mind.
— Seizing the initiative of the windward position, but with due regard for the possibility that contact might be made with his fleet to leeward.

— Full understanding of motion and distance relationships at sea, resulting (remarkably) in the achievement of tactical surprise and concentration of force in full view of the enemy.
— Training the way he would fight and fighting the way he had trained.
— High risk if his foe had been strong and competent, but little risk in the event, because his foe's gunnery and control of formation were poor and known to be so.

The tactical objective of the Royal Navy was usually to destroy the enemy fleet, which fitted Britain's strategic objective. The maritime objective of one of Britain's main opponents, the French, was considerably more involved, and in general French fleets did not have the size or training for head-to-head decisive battle. France's tactics responded to her strategic objectives and relative force inferiority, and she often enjoyed success. The predictable consequence, however, was long-term corrosion of her fleet's tactics, competence, and will to fight.

3

Tactical Development in Peacetime, 1865–1914

The Golden Age of Tactical Thought

The period from 1865 to 1914 rivals even our present age for sweeping technological development in peacetime. Whereas the age of fighting sail offered ample opportunity to learn on the job, the latter half of the nineteenth century saw few fleet actions that would serve as test beds, being as it was the climax of the Pax Britannica during which the Royal Navy was successful in dominating all naval rivals. Questions of tactics, technology, and command could not be resolved by battle, and thus the implications of steam propulsion, its effects on maneuverability, countervailing advances in armor and armaments, torpedoes and big ships versus small, and toward the end, the wireless and aircraft, all became the subjects of great debate.

The period was also a golden age of tactical thought, without parallel before or since. New armament, armor, and modes of mobility were imagined to carry the most extravagant tactical implications. By the end of the century the weaker tactical concepts had generally been discarded and a compatible marriage of new tactics and new warships had emerged, so that in World War I there were few surprises either in naval tactics or in the performance of warships. Postwar criticisms of both tactics and leadership confirmed more than challenged prewar analyses. It was the triumph of much hard thinking.

Rereading the tactical discourse of the period, one is struck by the tremendous thought and energy devoted to the application of mathematics to tactics. The logic of engineering skill spilled over from the design of war machines and influenced tactics tremendously. One cannot read Ambroise Baudry, Bradley Fiske, Romeo Bernotti, William Bainbridge-Hoff, S. O. Makaroff, and the naval journals and proceedings at the turn of the century without being inspired by the tremendous outpouring of technical and tactical creativity.* True, there were surprises, but compared with some of the absurd speculation that marked the early years of the technological transition, tactical analysis failed in two significant respects only: overvaluation of speed, and failure to foresee the effects that poor visibility would have on major fleet actions. The wisest writers wore naval uniforms, and they quickly tempered the more extreme tactical concepts that derived from theory but could not be implemented.

Precursors

When the Industrial Revolution blossomed after the Napoleonic Wars the effects were quickly felt in the navies of Europe. With the exception of modern methods of fire control and the automotive torpedo, all of the elements for the transition from sail to steam warships were conceived between the Napoleonic Wars and the American Civil War—steam propulsion and screw propellers; iron hulls and armor; bigger guns with greater muzzle velocity and penetrating power (ini-

*Some of the notable Naval Institute *Proceedings* prize essays of this period are Lieutenant Commander Richard Wainwright, "Tactical Problems in Naval Warfare" (Jan 1895); Lieutenant R. A. Niblack, "The Tactics of Ships in the Line of Battle" (Jan 1896); Lieutenant R. H. Jackson, "Torpedo Craft: Types and Employment" (Jan 1900); Professor R. Alger, "Gunnery in Our Navy" (Jan 1903); Commander Bradley Fiske, "American Naval Policy" (Jan 1905); and Lieutenant W. S. Pye, Jr., "The Elements of Fleet Tactics" (Jan 1906). Reynolds, whose characterization of the period as a golden age is more sweeping than mine, highlights strategy. He points out that the Russians were the first to publish a professional naval journal, the fine *Morskoi Sbornik*, dating from 1848, and that Italy's *Rivista Marittima* came to be the best of all the journals in the years before World War I.

tially these guns had more range, growing larger to smash through armor, but the range was not much more effective); breech-loading guns; effective shells and their necessary companions, fusing and rifled gunbarrels; and gun turrets.

There ensued the usual debate over the pace at which these inventions should be adopted, but it was evident that they would make the wooden ship of the line and its line of battle obsolete in short order. The British, slow as they were to adopt iron hulls, never built another wooden-hulled ship after they constructed the formidable nine-thousand-ton HMS *Warrior* in 1860, and the Royal Navy had to be rebuilt from scratch. Strategy too was overturned. Steam propulsion spawned a worldwide race for coaling stations, both necessitated by and impelling the spread of colonialism. Fortified ports were no longer adequate as naval bases; the need for repair facilities for guns and engines would diminish the feasible number of fleet sites. The shift from sail to steam would limit the range and endurance of battle fleets and ultimately have a profound effect on the form of blockade. None of these strategic limitations, which delayed the transition from sail, went unnoticed by the admiralties of Europe. The "modern" British navy was still building new ships with sails in the 1880s for strategic mobility.

On the other hand, the new tactical freedom brought by steam propulsion had tacticians fairly aglow with anticipation. Not only could a superior fleet attack directly into the wind, but it could close an enemy in the lightest wind and run the gauntlet of enemy fire at double or triple the former speed. While the full measure of this advantage was not, perhaps, appreciated until after the Battle of Lissa in 1866, the tactical discussions of the 1840s were already about the end of the coterminous column and the potency of ram bows.* At this stage armor was ahead of armaments in the technological race for superiority, so the attractiveness of the ram was linked to tactical mobility, kinetic energy, and the ability to close an enemy whose effective range and rate of fire had not kept pace with ship speed.

One combat laboratory for testing the rudimentary technology was the Crimean War. Although in that war logistics and strategic mo-

*See Robison, pp. 579–90.

bility, not tactics, drew preeminent consideration, the participants were convinced of the value of steam power for close-in work. Ironclads, not very seaworthy, nonetheless demonstrated their ability to stand up to forts and land batteries, which hastened their development in the Civil War. In 1853 the Battle of Sinope was seen as the proving ground for the explosive shell. Six large Russian ships descended out of the haze on seven hapless Turkish frigates and smashed them all, killing or wounding nearly 3,000 Turks at a cost of only 266 Russians.* Three of the Russian ships of the line carried new "shell guns" into the battle, and the navies of Europe marveled at their effectiveness. The battle may have been as influential in fostering armor and the iron-hulled ship as it was in hastening the development of shells, perceived as the nemesis of wooden warships.

Still, we may wonder at the reaction and speculate how much the shell guns promoted a battle conclusion that was preordained without them. A simple Lanchester-like model shows an outcome almost as overwhelming. If we credit each of the six Russian ships of the line with an average broadside of fifty-five guns and both of the frigates with fifteen, and oppose them with the seven Turkish frigates' broadsides of fifteen each, then by the time the broadsides of the latter were reduced to zero, the eight Russian ships would lose only two guns each.† Such was the power of Russian numbers, independent of the supposed quality of their materiel. In addition, the Russians probably achieved some advantage from surprise.

After the Civil War, 1865–85

The Civil War was almost devoid of fleet-on-fleet battles. Nearly all fleet actions were what we call today projection operations—inshore work—undertaken to control seaports, harbors, and rivers. The military targets were forts. "Fleet" actions occurred when the Confederates supplemented batteries ashore with ironclads, ships that were in effect mobile forts, heavily armored and armed. These inshore Confederate warships were everywhere effective but nowhere suc-

*Woodward (1965), p. 99.

†The three ships carrying the sixty-eight-pound shell guns each were 120-gun three-deckers. For simplicity I ignore the fact that one Turkish frigate escaped.

cessful for long, because the Union could always concentrate overwhelming numbers against them. The Confederates had no hope of assembling a fleet to challenge the Union at sea; their aim could never be any more ambitious than to conduct *guerre de course* against Union shipping, break the Union blockade locally, and hope for the intervention of the Royal Navy.

Consequently, the Battle of Lissa in 1866 served as the sole reference point for the study of fleet action. The following tactical lessons were drawn from that battle:

— Steam propulsion gave the offensive fleet new options.
— The ram was an effective weapon of naval combat.
— The single column would fail as a method of concentrating firepower.

It is noteworthy how quickly naval tacticians fell to the study of hypothetical fleet actions. Writings of the 1870s and 1880s extolled the ram. The ascendancy of armor plate over gun shot and early shells was so fleeting that we are prone to make light of the ram. But for twenty-five years or so the mobility of steamships allowed them to "charge" (a common description at the time) through a gauntlet of effective fire that was short, only a half mile deep at most. A fleet of rams could run eight hundred yards in three minutes or less and (it was thought) devastate a column armed with guns. And it could steam right into the wind. The more a defensive column closed up to concentrate its fire, the more vulnerable it was to ramming. The bigger the fleet was in single column, the longer the column and the easier it was for a ramming fleet to concentrate on a single segment. Figure 3-1, a schematic of the Battle of Lissa, illustrates this.

After 1866 tactical debate threw the traditional column overboard, and chaos reigned amidst all manner of speculation about the best tactical course to take. Solutions broke the long column into small components. Some argued for short mutually supporting columns, some for short lines abreast, some for units of two to five mutually supporting warships that could turn together as the unit commander dictated and that looked like a World War II circular formation—or the British infantry square. A melee was the logical consequence of each scheme, and tacticians despaired of finding any tactics that would give the fleet commander with a well-drilled navy anything but the

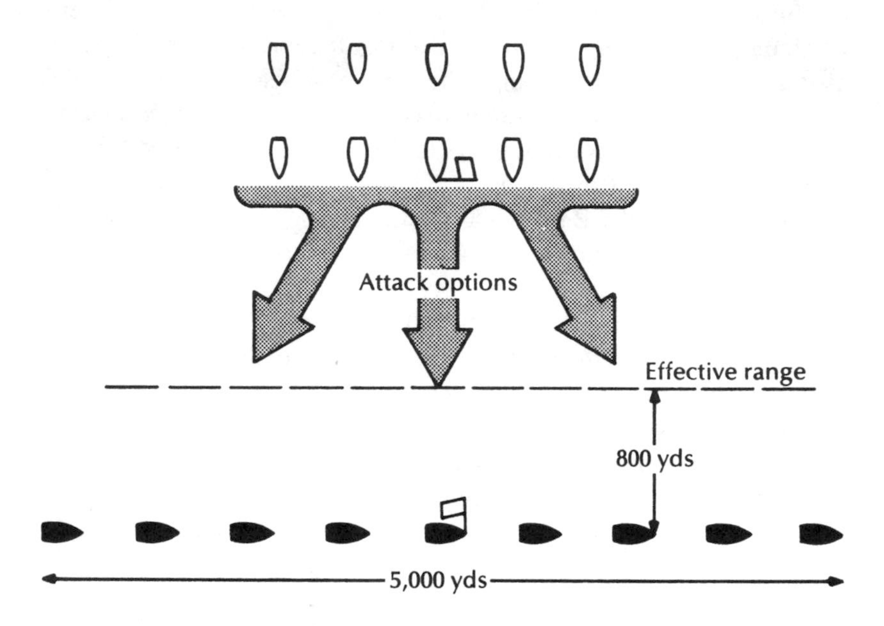

Figure 3-1. Employment of the Ram

This is a schematic of the Battle of Lissa, 20 July 1866. The Austrian victors are shown approaching in lines abreast against the Italian ships in column. Rear Admiral von Tegetthoff had only seven Austrian ironclads to face twelve Italian ships under Admiral Persano. Using the new freedom of maneuver afforded by steam, Tegetthoff steamed directly at the Italian column in V-shaped wedges, with his ironclads all in the first wedge. The Austrians smashed through the ineffective fire of the Italian column, throwing it in disorder and demoralizing Persano and his men. Only two ships were rammed, but the postmortems saw the threat of the ram in the Austrian charge as the key to victory, with lasting effects on warship design over the next thirty years.

most fleeting advantage during his approach. Many plans of action were based on strong analogies with land combat. Some were so amateurish as to envision the attacker maneuvering against a defending fleet that, like defending troops in ground battle, would be fixed in a stationary line.

Happily for the tactician, as well as the fleet commander who would have employed the tactics, technology came to the rescue before such a fleet battle was fought. No doubt it would have been chaotic. At best the new proposals were inadequate, but even at worst they did not lose sight of their aim, which was to concentrate all forces without a reserve, or the means to that aim, which was the newly available power of maneuver on the battlefield provided by steam propulsion. Nor did they forget the importance of the (fleetingly ascendant) ram. Signals and concepts were designed to give a commander an initial advantage; it was thought the battle itself would be shrouded in gunsmoke, determined partly by chance, and won, if at all, through the courage of individual captains.

Just before the end of the ram's speculative dominance, there were some engagements on the west coast of South America between 1877 and 1879 which, though hardly fleet actions, indicate what might have happened had major fleets engaged in battle. Four drawn-out battles between one or two ironclads and other warships resulted in many attempts to ram, nearly all unsuccessful. If there had been fleet actions, the results might have been more propitious for ramming, but the South American engagements showed that the difficulty of hitting a moving target had been underestimated.

A central participant in all four battles was the *Huascar*. She fought for Peruvian revolutionaries against Britain and then against Chile. She took an enormous number of hits at close range—sixty in her first battle—few of which penetrated, she participated in bombarding a fort, and she survived the war intact. The tacticians were right about one thing: the gun's effectiveness had been eclipsed, and for a brief period defensive technology was indeed dominant.

The Battle of the Yalu in September 1894 should have confirmed the end of ram tactics. The Chinese fought in line abreast with, according to a contemporary account, every intention of ramming with their two battleships, which were the heart of their fighting power.* But ramming played no part. The Japanese stayed in two columns and literally steamed circles around the Chinese. Firing range was very short. Most of the damage appears to have been done by

*Marble, pp. 479–99.

medium-caliber guns at ranges of about two thousand yards. There was an enormous number of rounds fired, and the cumulative effect of gunfire was devastating. The two Chinese battleships showed evidence of having taken 320 shell hits between them. Both, however, survived with their armor unpierced. Never did more heterogeneous forces fight with such different tactics, and never on a battlefield more difficult to analyze. Although the ram was unused, its utility was not laid to rest, and only in retrospect does the outcome of the great tactical debate start to take form amidst the clouds of intellectual ferment.

Triumph of the Big Gun, 1900–1916

It is widely thought now that gunnery usurped the ram, or that the ram was never an effective weapon at all. A better conclusion is that the torpedo superseded the ram. The Whitehead torpedo was a ram with reach: if it hit, it was almost as lethal and a lot safer to use. The study of gunnery became as obsessed with countering torpedo boats as with penetrating armor. Thus as the nineteenth century drew to a close, guns of many calibers, torpedoes, and rams were the weapons of interest, with the ram fast fading into oblivion.

The reemergence of guns as the central weapon at the beginning of the twentieth century and the confirmation of the armored battleship as the backbone of the fleet make for a complex, fascinating, and often-told story. Naval tacticians influenced the direction of the development. The Spanish-American War proved that even when gun shells had the potential to penetrate iron and steel plates, in at least two navies the gun could seldom hit a target in motion. Naval officers led by the brilliant Percy Scott and William S. Sims, motivators in the drive for gunnery accuracy, put their minds to improving fire control.* It was high time, because the thirty-knot torpedo boat and torpedo boat destroyer threatened to dash through the short gauntlet of secondary battery gunfire and wreak havoc on the battleships. In tacticians' estimates, a balanced fleet comprised more and more torpedo craft. Destroyers could in theory be forestalled with

*In one gunnery test observed by Assistant Secretary of the Navy Theodore Roosevelt, battleships fired two hundred shots at a condemned lightship at a range of 2,800 yards and obtained two hits (Mitchell, p. 148).

rapid-fire short-range weapons—effective torpedo range well into the twentieth century was a mere one thousand yards or less.* But the gunfire had to be accurate, and every commander at sea shuddered at the thought of having his battleship caught unprotected by his destroyer and light cruiser screen.

Here is how one tactician evaluated weapon effectiveness as a function of range in 1910:

Extreme range	10,000–8,000 meters	Heavy-caliber guns within range
Long range	8,000–5,000 "	Heavy- and medium-caliber guns effective, the latter against personnel and unarmored parts
Medium range	5,000–3,000 "	Medium guns worth a "special value"
Close range	3,000–2,000 "	Torpedoes a hazard, depending upon relative ship positions
Close quarters	Inside 2,000 "	Collisions possible (but no attention to ramming)†

The advocates of more and bigger guns in all-big-gun ships (the *Dreadnought* concept) could not yet make their case. Medium- and small-caliber guns had the same accuracy and higher rates of fire. Bradley Fiske presented three rules of thumb:
— A six-inch gun fired eight times as fast as a twelve-inch gun.
— A twelve-inch projectile carried eight times the energy of a six-inch-gun projectile.
— A twelve-inch-gun system weighed eight times as much as a six-inch gun.‡

Therefore on equal ship displacements, six-inch guns delivered eight times the projectile energy of twelve-inch guns. Muzzle energy, the product of projectile weight and muzzle velocity, was the quantitative

*See "The Destroyer Screen and Torpedo Threat," pp. 75–77.
†From Bernotti, p. 50.
‡Fiske, p. 25. Despite these data, Fiske became an advocate of the all-big-gun battleship with better fire control.

measure of a gun. Extensive computations of armor penetrability were made as a function of range, target angle, and plunging angle of shell. High rate of fire was also sought assiduously: a significant break point occurred at the six-inch gun, because its one-hundred-pound projectile was the heaviest that could be manhandled. In 1910 a six-inch gun had a nominal firing rate of twelve rounds per minute, which seems very fast indeed. If the big guns' long range could not be exploited with improved accuracy, then their greater ability to penetrate armor went for little. Small- and medium-caliber gunfire would inundate an enemy at short range, as the Japanese amply demonstrated at Tsushima in the Russo-Japanese War (1905), when the gunnery ranges that were entirely controlled by the Japanese were maintained at four to six thousand yards. To dominate the battle, big (ten- or twelve-inch) guns needed accurate fire control at ranges beyond the reach of medium (four- or six-inch) guns.

Still, the final outcome was predictable, if we recognize that one of the great trends in warfare is the increasing range of effective weapon delivery. In this century the change transpired with stunning speed. Around 1910, when continuous-aim fire and director control replaced local gunlaying, the all-big-gun ship was certain to dominate. Already the USS *Michigan* had been built, so close on the heels of HMS *Dreadnought* that American xenophobes were able to claim prior sponsorship of all-big-gun battleships, including a superior arrangement of turret.* All arguments fell into place for bigger and bigger battleships. Between 1905 (the USS *Michigan*) and 1912 (the USS *Pennsylvania*) displacements doubled. The details remaining to be settled were turret arrangements and the proper allocation of displacement between armor, speed, and endurance. Some extreme results were the battle cruisers HMS *Invincible* and *Repulse*, which had great firepower, high speed, and very little protection.

*Mitchell, p. 139, says dispensing with intermediate-caliber guns had been urged by progressive naval officers since 1901. Sims was the officer most effective in selling the all-big-gun battleship while destroying the contrary arguments of Mahan, who had lost touch with the fleet. For Sims' arguments, see "The Inherent Qualities of All-Big-Gun, One-Caliber Battleships of High Speed, Large Displacement, and Gun Power," U.S. Naval Institute *Proceedings* (Dec 1906): 1337–66. For a more popular perspective on guns and gunnery, there is also E. E. Morison, *Admiral Sims*.

By World War I, a mere ten years after Tsushima, big guns (twelve- to fifteen-inch) were *the* weapon, hitting repeatedly on a clear day after a few ranging salvoes out to eight miles and more. Behind the scenes and scarcely noticed, crucial developments were under way to improve the fire control computers that gave guns these long ranges. Secondary batteries of five- and six-inch guns were installed to ward off torpedo attacks, but every admiral wished to avoid having to use them for that purpose, because though the weapon debate was settled, the torpedo was still a threat to be reckoned with. In regards to this, the great Russian leader Admiral S. O. Makaroff wrote, with droll insight,

> Up to the present this [command of the sea] has been understood to mean that the fleet commanding the sea openly plies upon it and the beaten antagonist does not dare to leave his ports. Would this be so today? Instructions bearing on the subject counsel the victor to avoid night attack from the torpedo-boats of his antagonist. . . . Some seamen have become reconciled to this abnormality, yet if the matter were represented to a stranger he would be astonished. He would probably ask whether he properly understood that a victorious fleet should protect itself from the remnant of a vanquished enemy.*

Makaroff the tactician was explicitly challenging the authority of Mahan and Corbett, the strategists. The gun was in thought and fact the principal naval weapon in a fleet action, but the more perceptive tacticians were aware that the strategist's overly neat concept of command of the sea was rather too big a gulp to be swallowed whole.

Reemergence of the Fighting Column

The torpedo threat notwithstanding, all of the pieces of a tactical concept for fleet actions were in place and agreed upon. The ram was out of the picture. The battleship with the big gun was the decisive weapon. The bigger, the more stable, the better armed and armored the battleship, the better. It did not defend itself against torpedoes. Light cruisers and destroyers were built for that purpose; they would fend off enemy destroyers and torpedo boats. Scouting cruisers were the eyes of the fleet until airplanes or dirigibles offered assurance of

*Makaroff, p. 20.

greater competence. Battle cruisers, a last vestige of land combat's influence on naval thinking in the latter nineteenth century, would be a heavy cavalry in support of the scouting cruisers and able in theory to outrange or outrun any opposition. Mines were a wicked and ungentlemanly threat in shallow water but largely defensive, having to be planted by surface ships. Submarines, like mines with a deep-water offensive potential, were worse, an instrument of the devil. The wireless radio was a new tool of command, above all useful tactically to speed the results of scouting.

The tactical formation was the column, now restored to respectability for reasons seen in figure 3-2. Taking fifteen thousand yards as the range of effective fire in good visibility, a column of, say, sixteen battleships could engage anything with every centerline gun bearing over the full length of its nine-thousand-yard-long column. More than half the guns would bear at least within thirty degrees of the beam of the first or last ship in column. For comparison, figure 3-2 depicts on the same scale the narrow ribbon of death for a column of thirty eighteenth-century ships of the line.

Basic (but by no means ultimate) force-on-force correlations of the day started with opposing columns abreast. At least three tactical writers, J. V. Chase, Bradley Fiske, and Ambroise Baudry, described the cumulative effects of a superior concentration of firepower in the following way.* Assume that opposing warships each have a staying power of twenty minutes under the enemy's unopposed effective fire. Let the offensive capacity remaining (surviving guns and fire-control systems) be proportionate to the remaining staying power (in modern parlance we are speaking in terms of firepower kills). If all other things are equal, the battle will be a draw. It will also last a very long time because both sides' offensive capacities are reduced simultaneously.

*Anticipating Frederick W. Lanchester and his celebrated equations, the three naval writers explained their analyses using discrete time periods, which may be thought of as the time to exchange salvos. The engineer Lanchester used coupled differential equations, which were a cleaner and quicker way of making essentially the same points. Today the modern digital computer operates on discrete time differentials, the periods of which may be as small as desired. The authors wanted to illustrate that firepower effectiveness was not in simple direct proportion to weapon effectiveness (as, for instance, accuracy or rate of fire) but in proportion to the square of the number of weapons engaged. Robison casually refers to the latter effect as the N-square law.

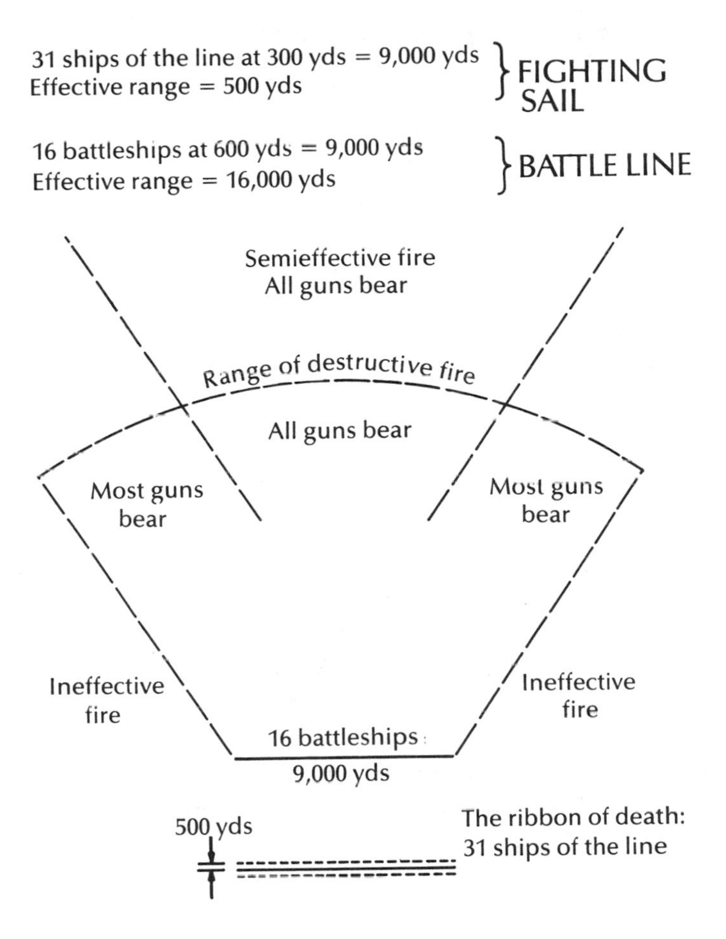

Fig. 3–2. Effective Range of Gunfire Compared, 1750 and 1910

But let side *A* fire for four minutes in advance of side *B*, as Baudry does,* and then table 3-1, of remaining firepower and staying power for both sides, shows that what had been an even battle now leads to the destruction of the side that was four minutes slow to open fire; being late, it lost (a seemingly modest) twenty percent of its fighting power. Side *A* takes twenty-six minutes to reduce the tardy side to impotence. That is only six minutes longer than it would take if *B*

*Baudry, pp. 116–17.

Table 3-1. Remaining Firepower and Staying Power

End of Minute	Units of Residual Firepower and Staying Power: Side A	Side B
0	10.00	10.00
2	10.00	9.00
4	10.00	8.00
6	9.20	7.00
8	8.50	6.08
10	7.89	5.23
12	7.37	4.44
14	6.93	3.70
16	6.56	3.01
18	6.26	2.35
20	6.00	1.72
22	5.83	1.12
24	5.72	0.54
26	5.67	0.00

had not returned fire at all. Also observe that the winner retains fifty-seven percent of his fighting power at the end of the battle.

Fiske built similarly simple tables to show the cumulative effect of preponderance of force. Let *A* now have two warships to concentrate on one of *B*. Under the same conditions of firepower and staying power as before, the table of surviving fighting power looks like this:

Table 3-2. Surviving Firepower

End of Minute	Superior Force A: Ship A_1	Ship A_2	$A_1 + A_2$	Force B: Ship B	Ratio of Fighting Values*
0.00	10.00	10.00	20.00	10.00	4.0
2.00	9.50	9.50	19.00	8.00	5.6
4.00	9.10	9.10	18.20	6.10	8.9
6.00	8.79	8.79	17.58	4.28	16.9
8.00	8.58	8.58	17.16	2.52	46.0
10.00	8.45	8.45	16.90	0.80	446.0
11.00	8.25	8.25	16.50	0.00	—

*Fighting values are defined as the square of the fighting power, and they indicate just that: the force's relative fighting value when concentrated.

Having been introduced to the Lanchester form of these calculations in chapter 2, the reader should not be surprised that after the inferior force is reduced to impotence, the superior force still retains 16.5 units (83 percent) of its strength. If a "continuous-fire" Lanchester form is used, the stronger force's surviving fighting power is slightly greater, 17.3 instead of 16.5. The difference is that, in the salvo model above, the weaker force's attrition does not begin until after it has fired without damage for two minutes.

All four of these theorists were successful practical men. Chase, Fiske, and Baudry were naval officers, and Lanchester was a prominent automotive engineer. They were working out different combinations of effects *for a situation in which the conditions held*. Fiske described quantitatively such things as the effect of more armor (a ten percent increase in firepower is worth more than a ten percent increase in staying power), better fire control, and a smaller effective target area (to compensate for a firepower inferiority of fifty percent, the effective number of hits taken must be reduced by seventy-five percent), and whether concentrating first on the greater or lesser force affects the theoretical battle outcome (it does not).

Fiske regarded such tactical parameters as the proper basis for determining ship design. The year the all-big-gun USS *Michigan* was authorized, 1905, he calculated his way to the conclusion that such a ship was unequivocally the battleship of the future. While Mahan was saying there were few principles of tactics because technology would always be changing them, Fiske was using abstract tactical models to guide the adaptation of technology and improve tactics.

Crossing the T

The similarity in the form of equations used in the age of fighting sail and the age of the big gun can conceal the major change in the way concentration of force was achieved. Although the column was the admiral's tactical formation during both periods, in sailing ships firepower had to be concentrated in the ship because gun range was so short, while in battleships firepower of an entire column—the firepower of every ship—could be concentrated. When the big gun dominated, it was weapon range that made "capping the T" so advantageous; instead of a single ship of the line in raking position, the whole fleet could concentrate fire on the enemy van.

The column was finally settled on (unequivocally after Tsushima) as the best offensive formation for simplicity of control and effective concentration of gunfire. As everyone knew, the column was vulnerable ahead or astern, and because of its motion especially ahead. Tactical discussions centered on how to cap the T. In tabletop tactical studies, the only means of achieving this concentration by maneuver was speed, and speed along with armor and armament was in every tactical and technological discussion.* As events transpired, the tabletop would prove to be misleading.

The decision in favor of the column simplified tactical thinking, and until World War II—which ended the relevance of the battle line—tactical discussions centered on:

— How to distribute gunfire: doing both what was best in theory (e.g., leaving no ship free from fire) and tackling the knottier problem of how to achieve the distribution. The fleet commander would be unable to *order* it in battle; the method had to be a part of doctrine or communicated with the simplest of signals. In practice, undistributed fire was to be a major tactical defect until the end of the surface gunnery era.
— How to cross the T against a manuevering enemy, and whether partial success had much value.
— How to achieve or forestall torpedo attacks by a swarm of destroyers.
— Where in the line to place the flagships.
— How to shift from cruising formation to battle formation.
— The new importance of scouting and measures to safeguard the scouting line.

Cruising Formation and Tactical Scouting

Greatly increasing weapon range and effectiveness were also having a profound influence on the need for reconnaissance. The commander of a large fleet in World War I had to have information about the enemy's force well before he could see it. Nelson's plan for Trafalgar,

*The U.S. Navy, after a lengthy and lively debate much influenced by Mahan, opted for heavy armor and armament at the expense of speed.

according to which the order of sailing was the order of battle, had lost its feasibility. For cohesion, communication, station keeping under cruising conditions, antisubmarine considerations, and rapid deployment in any direction, the cruising formation needed to be a series of short columns, each abreast of the others. Shifting to a single column for battle was a major and irrevocable commitment. The distance between columns had to be such that each column fitted exactly into the battle line, as depicted in figure 3-3. Battleship divisions were organized with the final sequence of ships in mind. It was important to maintain a close interval before and after the maneuver. All possible speed was also important, relatively more so than in the age of sail, when fighting sail meant minimum sail for steerageway. Still, the new battle fleet speed could be no more than two or three knots less than the slowest ships in the column, about a twenty percent margin, to allow the rearward ships a chance to keep station. It was observed that both seamanship and drills were imperative if the line was to be closed up and ships were to avoid colliding and masking one another's batteries. Although the column was the simplest formation, forming it quickly, keeping it closed up, and orienting it properly with respect to the enemy, the wind (gunsmoke was blinding), and the seas (rolling ships could not lay guns well) were skills derived from a lifetime at sea.

What looked simple on paper required consummate skill on the

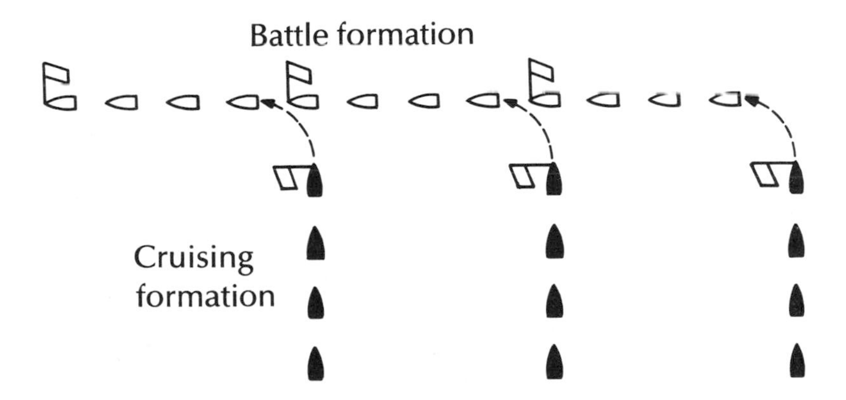

Fig. 3-3. Deployment of the Battle Line from Cruising to Battle formation.

ocean. Captains and flag officers received approbation or opprobrium based on their proficiency in close-order drills. One of the most famous of all maritime disasters, and a wonder of its day, was the collision in 1893 between HMS *Victoria* and HMS *Camperdown*, flagships of the Royal Navy's Mediterranean squadron. It occurred because the officer in tactical command, famous for tactical infallibility, ordered an impossible maneuver that went unchallenged.

The scouting plan received much emphasis, and considerable fleet resources were devoted to it. At Jutland both the British Grand Fleet and the German High Seas Fleet committed twenty to twenty-five percent of their heavy firepower and thirty-five to forty-five percent of their supporting cruisers and destroyers to scouting forces. The disposition is shown in figure 3-4. To give ships in the scouting line a measure of safety, they were permitted to fall back on the fast battle cruiser support group when the enemy was sighted and threatening. The whole of the scouting force would attempt to join the main body and augment it as the battle line went into action. But no one was sure that this could actually be done. The scouting line, which covered about thirty-five degrees on either bow of the battle fleet, was sufficient to sweep out a wide swath of ocean and protect against an undetected approach. No enemy could end-run the scouting line as long as the fleet was steaming ahead smartly. Reorienting the axis of advance of such a disposition must have been a tense and prolonged experience for every flag officer and ship captain.*

Command and Control

Increasing weapon range and gunnery effectiveness also demanded reconsideration of the position of the flagship. A battle would be decided very quickly once effective gunnery range was reached. Though it seems unlikely to the layman who thinks of ships plodding through the water at seventeen knots, there was no margin for error or delay

*Already aircraft (airplanes and dirigibles) were getting wireless radios and were seen as the scouts of the future. But, justifiably, they were not yet relied on. At Jutland Beatty launched a scout from the seaplane carrier *Engadine* immediately after the first scout cruiser's contact. The flimsy seaplane flew off in the wrong direction, eventually found some of Hipper's ships as it was flying beneath the low overcast, and reported what it could see to the *Engadine*. But nothing further came of this first tactical experiment with aerial naval reconnaissance.

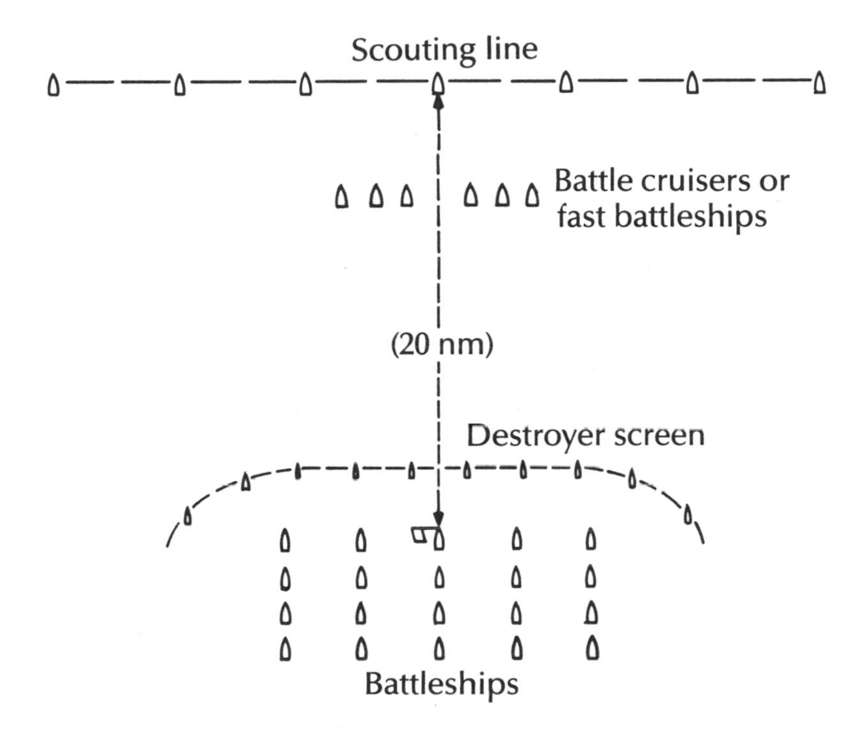

Fig. 3-4. Cruising Disposition, Showing the Scouting Line and Support Force

in maneuver. After experimentation and observation of the battles of the Yalu and Tsushima, the sensible conclusion was that the flagship should be in the van so that the commander could maneuver a column without signals, a simple "follow me" procedure. In contrast, turns together had to be executed precisely and therefore signaled, which took time. Simplicity and speed were everything. In the battle cruiser chase and counterchase that preceded the Battle of Jutland it took only a couple of minutes' delay in the receipt of a signal from David Beatty to Hugh Evan-Thomas for the latter to fall hopelessly out of the race while the former was losing two battle cruisers.

Still, Admirals John Jellicoe and Reinhard Scheer at Jutland had battle lines of twenty-eight and twenty-two battleships, respectively, which were simply too long to lead from the van. A fleet commander in the leading ship could not hope to know what was happening at the rear of his column six or eight miles away. He was better placed in the center: that position, though it was held at the expense of maneuverability, was the best for maintaining cohesiveness and control and visualizing the scene of action. Jellicoe knew that, if concentrated, his numerical advantage in firepower should compensate for any temporary loss of positional advantage. Scheer's concept of fleet management was similar: he had expected to entrap a weaker British element at Jutland. He had no desire to fight the entire Grand Fleet.

The dilemma of the commander of a large fleet is illustrated by Scheer's tactical position at Jutland: if he had been at the head of his column he might have foreseen and avoided the extreme misadventure of twice having his T capped. On the other hand, if he had been in the van he might not have succeeded in executing the famous simultaneous turns that allowed him to disappear into the haze two times and extricate his fleet from disaster.

Much progress had been made in formulating commands. Signals were not, as in the days of sail, keyed to the fighting instructions, the combat doctrine itself. Signals were the commander's way of communicating what was in his mind. Through continuing practice the signal book was to become a compact, unambiguous vehicle by which desired actions could be transmitted precisely. That this was so is inferred from the absence of discussion about the *system* of signals. By the end of World War II the signal book used by the U.S. Navy was a tactical instrument of collective genius, as reliable and thoroughly tested as the laws of physics. It was a treasure of efficiency, conciseness, and clarity—one that needs to be rediscovered by naval commanders at all levels today.

The signal book was only one component of the tactical communications system. Radio communications had arrived. The wireless made scouting lines feasible and, it is not too much to say, the cruising and battle dispositions of fleets themselves. Already the wireless was the subject of intercept and traffic analysis. Radio communications were one way the British and German navies sought to deceive and entrap each other. Early in World War I the British received a German code book recovered by the Russians from the hulk of a grounded German cruiser in the Baltic. It played a key role in leading Beatty's force of five battle cruisers on to a weaker force of four German battle cruisers under Hipper. That, the action off the Dogger Bank in January 1915, might have resulted in a crushing defeat of Hipper's force had it not been for two confusing tactical signals by Beatty. Signals warfare and cryptology had arrived. Both were being used tactically in ways that had never been seen before.

History books and old sailor's tales are full of lost, delayed, and misunderstood communications. What should we think of this? As far as tactical theory is concerned, the following: first, that the intent of the commander can best be learned by experience (teamwork is the result of a lot of work as a unit); second, that a set of signals, navywide and well practiced, is the next best way to suppress ambiguities and misunderstandings; third, that messages will continue to be lost, delayed, and misunderstood (no human system can eliminate communication errors—they are to be expected and insofar as possible hedged against in tactical doctrine); fourth, that the more one plans in advance, doctrinally and operationally, and the simpler one's plan, the fewer will be the communications and therefore errors in action.

This having been said, it is remarkable how few tactical communication errors occurred in World War I. A useful study would be to compare the total volume of tactical signals transmitted with the number of vital signals (those of grave consequence) and of signals that were in fact lost, delayed, or misunderstood. In World War I the number, as a percentage of the whole, was probably microscopic, serious though the recorded consequences of communication failures such as Beatty's were.

The Destroyer Screen and Torpedo Threat

The planned stations for destroyers and flotilla leaders were generally ahead of the van and abaft the rear on the engaged side.* The purpose

*For his book about the Solomons fighting, *Night Work*, Fletcher Pratt drew little diagrams of Japanese and American cruiser destroyer formations with the destroyers forming such bow screens. He was a great student of tactics before World War II, and he and his friends played his naval war game on his New York apartment floor. But sad to relate, his sketches rarely resemble the facts. Although his book paints vivid pictures of what command in the dark of night was like in 1942 and 1943, it is useless on the subject of tactics.

of these ships was to influence the main action: to force the enemy to turn or, if he did not, to cripple him with torpedoes. In battle, destroyers were treated as mad dogs on the leash of the destroyer flotilla commander. The fleet commander's practical control of them was limited to commands like "go" and "come." Their role was to rush in a tight pack and seize a battleship's throat if they could or, as was more likely, to leap and claw and growl at the enemy's own mad dogs, which had also charged into the fray at a single word from their master. If a destroyer was impaled on a battleship's bow and sunk, it was a scarcely mourned impediment to the battleship's effectiveness. If one was caught in the crossfire between battlelines, it might as well have been invisible. But a squadron of destroyers, bows on with a bone in its teeth, was a very visible and chilling threat indeed. A destroyer squadron commander fought for a semblance of order, living in the chaos of neglect by friends and spreading chaos among enemies as he was able.

The peacetime tactician's picture of the destroyers is represented in figure 3-5, the wartime tactician's picture in figure 3-6. Even the latter is really more like a spectator's view from the grandstand than a player's view, which is besmeared with smoke, confusion, noise, and fear.

Fig. 3-5. Peacetime Plans: A Battle Formation

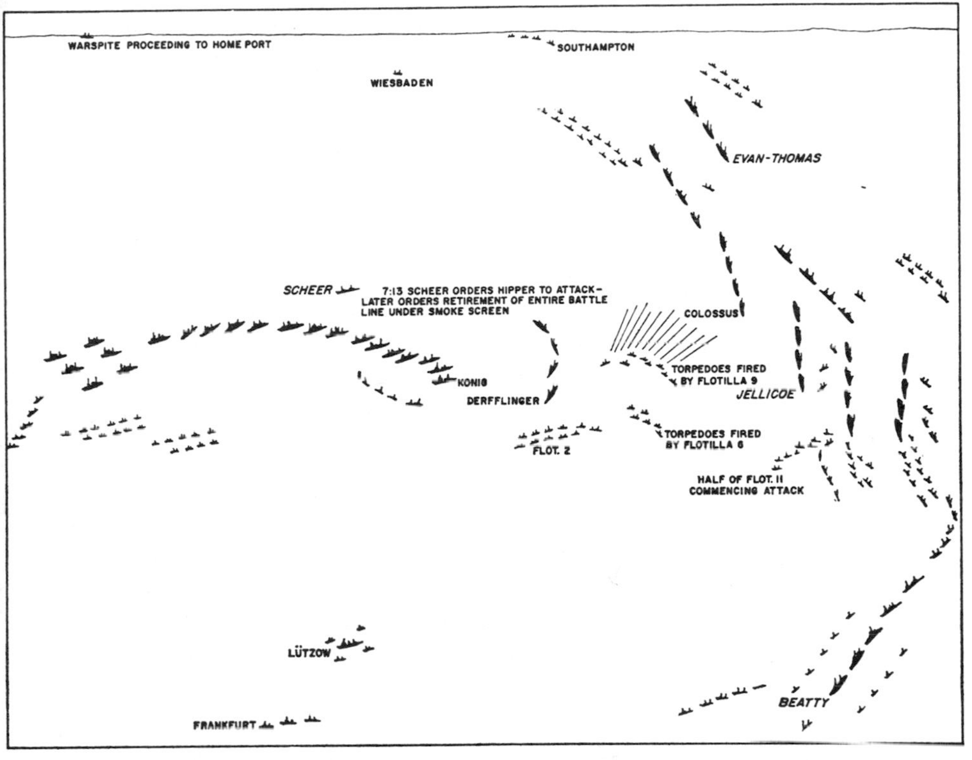

Fig. 3-6. Wartime Reality: The Battle of Jutland, Crisis at 1915

Turn-of-the-Century Theory and Practice

If all this disorder was inevitable, what good were precise mathematical calculations about weapon effectiveness? The answer hinged on what the fleet commander needed to know, which were some rough relationships between range and hitting effectiveness (e.g., when a torpedo-launching vehicle was a threat to the force) and *how the range was derived*. Some of the finest analytical thinking of the time on this subject seems to have been done at the Italian naval academy, judging from the writing of Romeo Bernotti and Guiseppe Fioravanzo. Bernotti's calculations of torpedo effectiveness in battle are a good single example of how astutely tactical quantitative theory and practical considerations were blended from 1890 to 1915. Bernotti treated in detail a thirty-one-knot torpedo with a running range of sixty-five hundred meters and a maximum speed of fifty knots.* For the purpose of dispensing with the simple-minded notion that raw running range was a significant tactical parameter, Bernotti ran through twelve pages of precise, concise, analytical, geometric, and probabilistic calculations of torpedo effectiveness, footnoting as he was able with (Russian) experimental results. Having calculated the effective range of a single torpedo against a nonmaneuvering two-hundred-meter target from different directions, he showed the mathematical advantage of firing from off the target's bow. Next he calculated the threat of a spread of five aimed torpedoes against a single target and of an unaimed spread against a battle line. He concluded that outside of thirty-five hundred meters (half the running range), though there was a perceptible risk, "from the point of view of the defense, there is no occasion to trouble oneself very much about it; and from the offense, it is well not to sacrifice, even to a minimum degree, the [effective] employment of the gun" (p. 25). On torpedo boat tactics, Bernotti argued for attacks by successive squadrons of three and demonstrated both the power of simultaneous torpedo launch and how to achieve it. He conceded that a coordinated attack is difficult in battle but showed that single-ship attacks in sequence brought with them scant probability of individual success. Bernotti's derivation of what seems today like point-blank range stemmed from a tactical philosophy that returns to haunt every commander planning a modern missile attack: "A weapon, the action of which cannot be repeated except at considerable intervals of time, and of which the supply is very limited, must be employed only under conditions that assure notable probability of hitting" (p. 14). Today's missile battle will center on keeping the enemy uncertain of his target and its position. Once launched, missiles cannot be recalled, and empty magazines can quickly become a terrible reality.

Will effective missile-firing ranges be shorter than their maximum ranges? Finding out is essential, judging from the Israeli and Egyptian

*Bernotti, pp. 13–25 and 161–71. Lieutenant Bernotti was then an instructor at the Royal Italian Naval Academy.

experience in the 1973 war. The missiles of the Egyptian warships outranged those of the Israeli warships. But the Israelis induced the Egyptians to fire all their missiles ineffectually, then closed in for a devastating finale.

Jutland

Like the combat data of all complicated battles, that of Jutland is difficult to assess. Ranges, gunnery accuracies, relative motion and position, speed of communication transmission and reception, and timeliness of execution—the meat and potatoes of tacticians—all are hard to trace among 250 warships. That sort of data, when it comes from smaller actions, is more easily sifted and compared with the data from controlled tests and experiments.

But there has been no shortage of second-guessing Jellicoe, the British tactical commander. A river of ink has been undammed in analyses of the battle he fought at Jutland. It was not only the last big battle of a war, like Trafalgar, but it also might have been, under slightly altered circumstances, the decisive battle envisioned by Mahan and Sir Julian Corbett. Therefore it was the centerpiece of analysis in all navies until World War II. Jellicoe did not succeed in destroying the German High Seas Fleet, and the postmortems had to decide the reason. He failed where Nelson had triumphed. The controversy boiled down to why Jellicoe, with a superior fleet which had taken Scheer by surprise, had not been more aggressive. His famous turn away from the desperate German destroyer attack, his tenacious faith in the single column, and his other conservative measures may all be explained on purely tactical grounds—the self-imposed demand for control. But the heart of the matter was that battle tactics on both sides were governed by the two national maritime strategies.

It does not do simply to dismiss Jellicoe as lacking the Nelsonian will to win. The analyses that reach this conclusion also compare his fleet's quality and gunnery unfavorably with the German fleet's, some authors going so far as to say Scheer had a chance to win.* One cannot have it both ways. Either Jellicoe retained sole power to

*For example, see Fioravanzo, p. 154, or Hough, *Great War*, p. 122.

destroy, provided he exercised no imprudence, or his caution was justified because he in fact might have lost, in which case the consequences were incalculable. There is no question that Jellicoe had the forces to win. Scheer fled the field. But at what odds? That is the key question.

If we accept for examination the common presumption that strategically it was more important for Jellicoe not to lose a decisive battle than to win one, then a reasonable tactical estimate is that he could not lose if and only if he maintained offensive concentration through fleet cohesion. His 3:2 numerical advantage in dreadnoughts guaranteed that. Better German gunnery and ship staying power could not redress the numerical advantage of the Grand Fleet. However, with quality taken into account, the force-on-force comparison might have been as narrow as 4:3 or even 5:4. That would have been close enough to permit the High Seas Fleet to win with a brief ten-minute advantage in concentrated fire if, for instance, the tactical positions had been reversed and it had been the British fleet's T that was capped.

With the fleets positioned as they were on that late May afternoon, Jellicoe's biggest danger lay in an effective enemy destroyer attack. Perhaps the danger was less severe than Jellicoe envisioned it.* But we must remember that he had announced what he would do in the circumstances, not only to the Admiralty but also to his captains, and he had to fight his fleet as he had trained it. A signal to turn toward rather than away from the torpedoes at that moment was an invitation to pandemonium. If and when he ordered a turn, Jellicoe was bound to turn away.

As far as an attack by the British destroyers was concerned, no such opportunity came to Jellicoe. His best tactic was to use his battleships for firepower and his destroyers to screen. A destroyer attack is an equalizing tactic, one to be risked by the weaker force, and his fleet was not that. Moreover, his tactical position was ideal for the battle line and questionable for a destroyer attack.

*I have revised my opinion and believe now, contrary to H. H. Frost and many other critics, that Jellicoe's estimate was accurate. This conclusion resulted from study of the devastating effects of large torpedo spreads in the night actions around Guadalcanal. See pp. 117–29.

There was no tactical initiative open to Jellicoe that would have been consistent with the offensive spirit of Nelson. It is difficult to imagine one even in theory, given the weapon characteristics of 1916. When Beatty assumed command from Jellicoe eighteen months after the battle, all possible lessons had been assimilated. Yet the changes in tactics introduced by that colorful admiral were insignificant. Nor do we see any Nelsonian changes to fighting by squadrons or anything at all radical in the battle tactics of the 1920s and 1930s. And as for the execution of some theoretical tactical initiative on the spot, it would have been impossible. Only today's game players, who have the magical power to move whole fleets on video screens with buttons, can carry out such unpracticed maneuvers. An admiral of a fleet does not expect to exploit opportunities with tactics that have not been inculcated.

It is reasonable to believe Jellicoe (and Scheer) knew this. If so, the explanation for Jellicoe's actions is one he could never offer in public: that, ship for ship, Scheer's were better, and Jellicoe's superiority rested narrowly on maintaining unity and mass. Any concentration without massing was unmanageable. Jellicoe could not fight like Nelson because he was not fighting a fleet like Villeneuve's. Jellicoe was opposed by superb weapon systems, skillfully handled.

Concept Versus Reality

If tactical theorists had underestimated the smoke and confusion of big battle, it is evident from his handling of Jutland that Jellicoe had not. Tactically he executed what he conceived to be his mission: to bottle up the High Seas Fleet, make his numbers count, win as he could, and avoid loss due to carelessness, enemy wit, or bad luck. His so-called mistake, a preference for risk avoidance over aggressiveness, was a calculated, preprogrammed, doctrinal, and totally predictable appreciation of Britain's maritime strategy. In fact, the refusal of the High Seas Fleet to come out and submit to its calculated fate was only the first of several unanticipated *strategic* surprises; equally significant were the threat of the submarine to the British sea lines of communication, the snail-paced impact of the strategic blockade, and the effect of mine warfare on all kinds of naval operations.

The new weapons changed not only tactics but strategic and logistic plans as well.

Tactical plans served well, owing to the great deal of preliminary thinking and writing that had been done and debate that had been carried on by the naval officers who executed them. A handful of tactical surprises, however, did arise. Prominent among them was an almost total disregard before World War I of the importance of deception. Every major engagement in the North Sea, the cockpit of the naval war, was part of an effort at seduction. Both sides knew the advantage of numbers and the N-square law. Neither fought voluntarily when outnumbered ever so slightly, so trap and counter-trap became the method of war. And more often than not the schemes backfired, even at the Battle of Jutland.

If planned surprise through traps did not work very well, unplanned surprise through failures in scouting abounded. Beatty, Jellicoe, Scheer, and Graf von Spee (at the Falklands) all suffered rude surprises by accidents stemming from scouting deficiencies. A dominant but strangely unpredicted feature of Jutland was poor visibility, caused by the gun and engine smoke of 250 warships. There is something about the game floor or video screen that deludes the tactical planner, who may forget that circumstances can drastically foreshorten opening ranges and change the whole nature of the battle. Certainly that is what happened to the Americans in the Solomons night actions in 1942–43 when, unlike the rehearsals of the 1930s, the battles often opened at point-blank range. The American navy of today, used to the wide sweep of the oceans, can easily forget that inshore work now extends many miles out of sight of land. Submarines aside, there may be many opportunities for hide-and-seek in the Mediterranean and Norwegian seas. If history is any guide, planned surprise can be achieved by a fleet, and unintended surprise is always near at hand.

Which brings us to speed. Prewar writers thought, correctly, that gunfire would work very quickly once forces were within effective range. Therefore they envisioned elaborate maneuvers beyond effective range to achieve an advantageous position. In practice, engineering speed used to achieve tactical advantage usually went for nought. The speed of a fleet was the speed of the slowest ship in formation. The lesson of the Japanese advantage at Tsushima was learned too well. The trend from the period of the ram through that of short-range gunnery and into that of director-controlled long-range gunnery was a steady erosion of the advantage predicted for warship speed. In addition, some prewar writers omitted from practical consideration the problem faced by the commander with damaged ships that must slow down. Mahan wisely said that "the true speed of war is . . . the unremitting energy which wastes no time."* Failures to consummate opportunities were failures to communicate and comprehend clearly. The speed that mattered was in the realm of decision making: decisions had to be quickly made and transformed into simple, correct maneuvers.

And what then of simplicity? When the chips were down the fleet commander opted for simple formations to maintain control. Echelons, boxes, anything that gave more theoretical superiority were set aside. To achieve his astute maneuvers Heihichiro Togo took the lead and guided his own column, using the simplest mode of control. We may believe he knew that by leading his single simple column he could make his twelve-ship battle line maneuver with one mind.

Summary

Until the start of World War I there were few opportunities to see the effects of the many technological advances in battle, and so the period between 1865 and 1916 is a case study of the relationship between theory and practice. It was a time of extravagant speculation, but it culminated in superb tactical thought, incorporating what today we call operations analysis. Most of the tactical hypotheses were advanced by naval officers and debated in professional journals.

These tactical studies succeeded admirably in preparing wartime leaders. The battle fleets fought as they had been expected to. The rigorous debate had not overlooked the importance of the column,

*Quoted in Hughes, p. 193. In that essay, the Mahan quote was preceded by the following: "The great end of a war fleet. . . is not to chase, nor to fly, but to control the seas. . . Not speed, but power of offensive action, is the dominant factor in war. . . Force does not exist for mobility, but mobility for force. It is of no use to get there first unless, when the enemy in turn arrives, you also have the most men, the greater force. . . The true speed of war is not headlong precipitancy, but the unremitting energy which wastes no time."

scouting, concentrated firepower, leadership, training, morale, or C^2. Only deception, accidental surprise, and the limited payoff of ship speed were underestimated or miscalled.

By the end of this period everything was aimed at concentration of long-range, big-gun firepower. Superior concentration was achieved through massing and maneuver. For most of the period fleets saw each other and maneuvered before they were within effective range, but by the beginning of World War I they needed to deploy for battle while still out of sight, and so tactical scouting assumed prominence and consumed resources.

Command was keyed to a faster pace. Simplicity and doctrine dominated tactical procedures.

Other trends emerged as well:

— An increase in the range and destructiveness of weapons
— Oscillation between the dominance of armaments and the dominance of armor
— An increase in the speed and maneuverability of weapon systems in battle (though tactical maneuverability afforded by steam propulsion came at the expense of *strategic* mobility)
— An increase in the importance of tactical scouting as weapon ranges and ship speeds increased

Most surprises, though generated by technology, were strategic. The tactical roles of guns, torpedoes, mines, and every class of ship had been anticipated. But either one or both sides failed to predict the end of close blockade, the strategic success of submarine warfare, and the pitfalls of amphibious warfare. All were major changes.

4

World War II: A Weaponry Revolution

Surprise or Upheaval?

The phenomenal shift in tactics during World War II took nearly everyone by surprise. Even those who professed to have foreseen the tactical revolution—the air power zealots—foretold too much too soon. Pearl Harbor and the Battle of the Coral Sea were the culmination of events that had been building through the peacetime 1920s and 1930s. As we discuss the tactics of carrier warfare it is worth remembering that everyone was learning on the job. One only needs to read Bernard Brodie's 1942 edition of *A Layman's Guide to Naval Strategy* to appreciate the turmoil in many minds three years after fighting broke out.*

The estimable Brodie need have had no cause for chagrin. An ex post facto reading of naval operations off the northern European coast and in the Mediterranean leads not to the conclusion that the ascendancy of air power should have been obvious but to an appre-

*See chapters 8 and 9 of that book. Brodie's *Sea Power in the Machine Age* (1943) is also insightful.

ciation of how close the competition between gun and airplane really was. For instance, in 1940 two German battleships caught the British carrier HMS *Glorious* in the open sea and sank it. By 1944 U.S. Fleet antiair-warfare (AAW) defenses were so impregnable that Japan had to abandon bombing attacks and resort to kamikaze missions. Land-based horizontal-bomber attacks against warships—one of the B-17's original missions—were not effective.* The torpedo bomber, while scoring successes, came to be a kind of unintentional kamikaze. In the end only the dive-bomber spelled the difference. As usual, vision played its part in the rise of naval air power, but it was the pragmatic tactician and technologist arm in arm who worked out the details.

The "battleship admirals" were not as important as they have been thought. For one thing, aerial bombing tests in the early 1920s against the old *Indiana*, *New Jersey*, and *Virginia*, and the new but uncompleted *Washington*, along with Billy Mitchell's rigged attacks on the *Ostfriesland*, proved not so much that heavy bombs could sink warships but that the aircraft of the day would have great difficulty sinking a moving, defended, buttoned-up warship. For another, the 1920s were a time when crucial decisions about development were being forcefully supported by the navy. Between 1922 and 1925 naval aviation's budget held steady at 14.5 million dollars while the navy budget as a whole shrank twenty-five percent. From 1923 to 1929 the naval air arm increased by 6,750 men, while navy manning overall decreased by 1,500—and this is excluding the crews of the manpower-intensive *Lexington* and *Saratoga*.†

In an astonishing sleight of hand, carrier tonnage was allowed among the five major signatories of the Washington Disarmament Treaty of 1921—the United States, Great Britain, Japan, France, and Italy—at levels of 135, 135, 81, 60, and 60 thousand tons respectively, at a time when "no naval power . . . possessed a single ship that could be applied against the allowed carrier tonnage. All carriers in service or building were to be classified as experimental

*It was easy to explain geometrically why. For an elementary analysis see Fioravanzo (1979), pp. 177–78. We should also add that high-level bombers coordinated with simultaneous low-level attacks could be effective against poorly defended merchant ships. The remarkable Battle of the Bismarck Sea proved that.

†Melhorn, pp. 93, 94, and 154.

and therefore did not count. . . . Drastic as were its reductions in capital ships, the conference clearly determined that there would be no statutory interference with the development of aircraft carriers."* From 1921 to 1935, the treaty years, there could be as much as but not more than about one-third as much carrier as battleship tonnage. Both Japan and the United States built every ton allowed for carriers. The Washington treaty and those that followed were not a constraint on but an incentive for air power.

William Sims, Bradley Fiske, William Moffett, Ernest King, Joseph Reeves, and Thomas Hart were among the American surface officers who appreciated very early the importance of naval air and encouraged aviators like Henry Mustin, Kenneth Whiting, John Towers, and Marc Mitscher to hasten its development. The United States led the way, eyeing the broad Pacific, followed closely by Japan. Britain's naval air arm, backward in some ways, would prove to be the best in the Mediterranean. Still, no naval power could predict the dominance of naval aircraft. Technology which lay like a sleeping giant between the wars was prodded awake in 1939 by combat. Even then the issue was in doubt. Consider one of Charles Allen's perceptive illustrations of the connection between technology and tactics:

> In the delicate balance of interactions it is noteworthy that the greatest swing factor in the battleship versus carrier issue may have been the actual performance of the newly introduced technology of radar. If it had proven more effective in directing heavy AA guns [or if, as others have said, the proximity fuze had come along a few years sooner], the effectiveness of tactical strike aircraft might have been largely neutralized. If it had been markedly less effective for early warning and fighter direction, carrier vulnerability might have been too great to bear. In either case, the fleet would have been dramatically different in 1945.†

That aircraft would have *some* vital role was foreseen by all the powers. Aircraft were essential as scouts and, not to be overlooked, acted as spotters for gunfire in those days before radar. They were useful enough that the battle force deplored the possibility of losing

*Melhorn, p. 83.
†Allen, p. 77.

its air cover. But if carriers were positioned too closely to the battle line, they would be exposed to attack. As early as 1930, commander, Aircraft Squadrons, Scouting Fleet, wrote, "Opposing carriers within a strategical area are like blindfolded men armed with daggers in a ring. There is apt to be sudden destruction to one or both."* In developing their interwar plans, inferior navies—the Japanese when fighting the United States, and the United States when fighting the British—expected to use aircraft to soften up and slow the enemy battle line. Whether air power should be land or sea based was debated everywhere, but the need to command the air space over the fleet was acknowledged by all but the least perceptive, and fighter aircraft were seen as key players in this. By the 1930s U.S. and Japanese carrier aviators knew their own potency and used every infrequent opportunity to experiment. The Japanese navy already practiced leading with carrier strikes; and for insight into what U.S. naval officers thought was the paramount threat, one has only to consider the intensity with which naval intelligence tried—and failed—to track not Japanese battleships but Japanese carriers just before the raid on Pearl Harbor.

But how would battles be fought and what would the tactics be? These were questions hardly answered in 1941, even after two years of fighting in the North Sea, the Atlantic, and the Mediterranean.

This situation is in striking contrast with that prior to World War I. Then technological advances had been assimilated and fleet tactics set in place. At the onset of World War II technology was in ferment and tactics had not yet caught up with their potential. The visions of those who thought aircraft would sweep the oceans clean of surface warships were so patently premature that they helped fuel the passions of the conservatives, who clung doggedly to the supremacy of the line of battleships. In large measure the fleet tactics anticipated for World War II echoed those for the previous war, except that aircraft, the new mad dogs of their day, would fight each other overhead. This was the situation in all navies. Tacticians had to adapt in the midst of the war so extensively that by the end of it no major

*Letter (May 1984) to author from Dr. Thomas C. Hone of the Naval War College faculty, who has studied the surviving documentation from the U.S. Navy fleet problems, 1929–39.

category of warship except minecraft was employed in the U.S. Navy tactically for the purpose for which it had been built. The striking and supporting roles of battleships and aircraft carriers were reversed; heavy cruisers, designed in part for fleet scouting, did nearly everything but that; light cruisers designed as destroyer leaders became AAW escorts for carriers; destroyers conceived for defending the van and rear of the battle line against torpedo attacks from other destroyers were adapted to function as ASW and AAW escorts; and submarines designed for forward reconnaissance and attacks on warships were diverted also to attack merchant ships and the sea lines of communication. By the end of World War II the upheaval of tactics, hastened by technology, was complete. Along the way tactical problems had to be solved.

In the fleet exercises of 1929 the *Saratoga* made a night run around the defending fleet and conducted a successful air strike against the Panama Canal. The attack is celebrated as the symbolic arrival of carrier aviation as a force to be reckoned with. But with it came double-edged ramifications. After launching the strike, the *Saratoga* was found and "sunk" not once but three times—by surface ships, a submarine, and aircraft from the *Lexington*. As it turned out, the center of concern in World War II was the vulnerability to naval aircraft of warships of every description, foremost among which was the carrier of the aircraft itself.

Five New Tactical Problems

Among the many tactical problems facing U.S. and Japanese naval commanders in the Pacific, five were prominent. Being interrelated, they were the more difficult to resolve.

1. *The tactical formation.* Thanks to prewar experimentation, the advantages of the circular formation for the defense of a carrier were understood by both U.S. and Japanese naval aviators. For the United States, the many-faceted radar made station-keeping easy. Offensively the formation could be maneuvered by simultaneous turns to maintain unity during flight operations under radio silence. Defensively a circle was best because it guarded against enemy aircraft seeing a gap in the screen of escorts and exploiting it. The question was whether each carrier should have its own screen to maximize its

flexibility or whether two or three carriers should be surrounded by a single, stronger ring of escorts. Protection against submarines was also a consideration. A "bent line" screen would do better for that purpose, but it was generally incompatible with carrier operations; speed, then, was the carriers' best security against torpedoes from slow-moving submerged diesel submarines, along with a policy to avoid steaming through the same waters repeatedly.

The effectiveness of air offense and defense was the issue. The Japanese established separate carrier formations at the outset and changed only when forced to by a shortage of escorts. In the U.S. Navy the issue was less clearcut: the battle between senior aviators began to peak after the Battle of the Eastern Solomons, when one commanding officer added a new wrinkle by contending that the *Saratoga* had escaped attack and survived because of her ten- or fifteen-mile separation from the *Enterprise*, which was heavily damaged. Was it not better to lose one carrier and save the other than to lose two carriers to a concentrated attack? Which should take precedence, passive defense through physical separation with concomitant flexibility for air operations, or better AAW defense through compact defense?

2. *Dispersal or massing?* The range of attack aircraft opened up the possibility of concentrating offensively from two or more carrier formations that were physically separated by hundreds of miles. In practice, the need for radio silence hampered—perhaps spoiled—this possibility, and the United States never entertained it. American tacticians argued over separate formations but they kept the formations close enough that the fighter air defense—the combat air patrol (CAP)—could protect the entire carrier force. For the U.S. Navy, concentration and massing were synonymous.

The Japanese had a penchant for dividing their carriers, and they have been much criticized for it. E. B. Potter, for one, takes Vice Admiral Takeo Takagi to task for hoping with his approach in the Battle of the Coral Sea to "catch the American carriers in a sort of pincer movement."* Later in 1942 Yamamoto's plan for the Eastern Solomons battle placed the light carrier *Ryujo* in front of the two big

*Potter, p. 664.

carriers as a decoy. She was sunk, and the Japanese have been condemned for dividing their forces. As explanation for complicated Japanese dispositions, U.S. critics have called Japan "sneaky" and pointed to her history of surprise attack. No doubt surprise attack—an effort to attack effectively first—was the basis of Japanese planning, but why divide carriers? At the Coral Sea the main striking fleet took advantage of a weather front and approached from a direction toward which land-based U.S. aircraft could not search and carrier aircraft were less likely to search. A pincer movement is an absurdity for someone as astute and familiar with carrier air power as Yamamoto. Should we not seek a better explanation for these strange Japanese dispositions? The answer is yes, and the explanation derives from Japanese faith in the dominance of successful air strikes.

3. *Offensive vs. defensive firepower.* Though the tactical commander must fight with the forces at his disposal, he has choices. He can emphasize fighter escort for his strikes, or he can emphasize his fighter CAP. He can add fighters for defense to his flight decks and carry fewer bombers and torpedo aircraft, or vice versa. He can use most of his scout bombers for scouting, or he can take calculated risks in scouting and husband them for a stronger attack. He can integrate his battleships in the carrier screen for AAW defense, as the U.S. Navy did in the Pacific, or he can keep his battleships separate for offensive follow-up attack, as the Japanese did. These decisions hang much on the estimate of the power of the offense. Clark G. Reynolds, like many commentators, is impatient with Raymond Spruance for failing to use his carriers more offensively in 1944.* Were not the enemy carriers Japan's threat and America's objective? Spruance was the best U.S. naval tactician in World War II. Then why didn't he act as the Japanese had, leaving a small force to cover the amphibious assault force and using his fast carriers to go after the Japanese carriers and hit them first?

4. *Daytime vs. nighttime tactics.* Carriers dominated the daylight hours but were sitting ducks for gunfire at night. Detaching before darkness, a battleship or heavy cruiser formation could travel two hundred nautical miles at night, a distance engraved in every tactical commander's mind. Since air strikes were mounted at ranges of around

*Reynolds, pp 181–205.

two hundred nautical miles, a carrier force could not be closed unless it pursued a crippled and presumably retreating enemy. Because of the damage to the U.S. battle line at Pearl Harbor, there would not need to be a command decision in 1942 about whether to send gunships against gunships halfway between two opposing carrier forces two hundred nautical miles apart.* But the Japanese, who were the aggressors in 1942, three times sent their surface ships carrier hunting. The U.S. tactical problem in 1942 was whether to pursue after dark and risk an encounter with Japanese gunships or stand clear and let the enemy warships or invasion force steam safely away. In 1944 the problem was whether to employ the fast battleships as a unit for offensive action in the Japanese fashion or keep them with the carriers for defense. What was the basis for tactical decision?

5. *Dual objectives.* According to typical American prewar tactical planning, the U.S. battle fleet steaming west to relieve Guam and the Philippines would be met by the Japanese battle fleet and a great decisive action would occur. It is true that as logistical considerations intruded, the simple tactical paradigm was complicated by the need for bases and the fleet train. But guarding the train or an invasion force was not yet a thing that fleet tacticians worried much about.

The airplane changed that. Until there was a threat of invasion by the navy on the strategic offensive, a weaker battle fleet on the defensive could not be induced to fight. But an invasion force had the responsibility of protecting amphibious assault ships, and with aircraft in the offing this presented new and complicated problems. Aircraft had to cover the transports as well as attack the enemy. In all six of the Pacific carrier battles the attacker had a primary or secondary mission to attack and destroy the enemy fleet. In each instance an amphibious operation was involved. Obviously the attacker did not want to expose his transports. This dual objective was inescapable for the Japanese in 1942 and for the Americans in 1944. Tactical plans and decisions also had to deal with the new problem that strategic offensive brought: how to dispose forces while protecting transports. In the era of aircraft, tactical commanders had to solve the unprecedented problem of enemy attack from a long range.

*Such a night action was a real possibility at the Battle of the Coral Sea, a remote possibility at the Battle of the Marianas. Of course, under rather different circumstances it occurred in the Battle for Leyte Gulf.

A Tactical Model of Carrier Warfare

The five major issues of Pacific carrier tactics can be illuminated in the context of a simple model, which will also promote understanding of the model of modern missile warfare to be discussed later. The model of carrier warfare compares with the Lanchester-Fiske model of gunnery in several ways. Fiske envisaged a mutual exchange of salvos that would erode the residual strengths of both sides simultaneously. His purpose was to show the cumulative effectiveness of superior firepower, the dominance of a small advantage if the advantage could be exploited with coherent maneuvers, and the disproportionately scanty damage the inferior force would inflict, no matter how well it was handled tactically. Gun range was a matter of indifference to Fiske because both sides faced the same range. He felt free to disregard for purposes of illustration the possibility that one side could outrange another and maintain an advantage that was in any way consequential. In effect, the pace of the battle would accelerate as the range closed, but the final ratio of losses would not change. His model took into account the "staying power"—warship survivability—that accorded with the assessments of his day: a modern battleship would be reduced to impotence in about twenty minutes by unopposed big guns within effective range.

The gunfire model of simultaneous erosive attrition does not work for the World War II carrier offensive force. That force is best represented as one large pulse of firepower unleashed upon the arrival of the air wing at the target. If, as was common, the second carrier force also located the first and launched its strike, simultaneous pulses of firepower would be delivered from both fleets. If the second carrier fleet did not find the first in time, it would have to accept the first blow. By then it would probably have located the first force and, if there were any attack capacity remaining, it would strike back.

To calculate damage from an air attack it is necessary to figure the defender's counterforce as the combination both of active defense (fighters and AAW strength) and passive defense (formation maneuverability and carrier survivability). In the Pacific, effective carrier-based air attack ranges were comparable, 200 to 250 nautical miles, and neither side could outrange the enemy's carrier aircraft. So in carrier battles, the crucial ingredients were *scouting effectiveness* and *net striking power*. Scouting effectiveness came from many sources: raw search capability, including organic and land-based air reconnaissance; submarine pickets; intelligence of every kind; all enemy efforts to evade detection; and, not to be overlooked, the planning skill of the commander and his staff. Net striking power was made up of raw numbers of attacking bombers and fighter escorts, reduced by the active and passive defenses and the relative quality of material and personnel on both sides.

For our purposes now, scouting effectiveness will be determined simply by asking who attacked first or whether the attacks were simultaneous. As for striking effectiveness—damage inflicted—the crucial question is the value of a carrier air wing's strike capacity. Of course there is much room to examine tradeoffs in practice between attack aircraft used for scouting or attacking and fighters used for escort or CAP. These were problems air staffs had to deal with. I need not introduce them in detail here.

For the moment I will assume (not unreasonably, as we will see) that in 1942 one air wing could on balance sink or inflict crippling damage on one carrier and that cumulative striking power was linear: two carriers were about twice as effective as one and so could sink or cripple two. A very rudimentary table of outcomes after the first strike can be constructed for three cases: (1) the equal or superior force *A* attacks first; (2) the inferior force *B* attacks first; or (3) *A* and *B* attack together.

Table 4-1. First Strike Survivors (A/B)*

	Initial Number of Carriers (A/B)				
	2/2	4/3	3/2	2/1	3/1
(1) A *strikes first*	2/0	4/0	3/0	2/0	3/0
(2) B *strikes first*	0/2	1/3	1/2	1/1	2/1
(3) A *and* B *strike simultaneously*	0/0	1/0	1/0	1/0	2/0

*It is immaterial here whether the nonsurvivors are sunk or out of action. But later we will take survivors to mean carriers with operational flight decks and viable air wings.

If we allow the survivors of the initially superior but surprised force *A* to counterattack, the final outcome is:

Initial force (A/B)	2/2	4/3	3/2	2/1	3/1
Survivors (A/B)	0/2	1/2	1/1	1/0	2/0

It may be inferred from reading the views of naval aviators at the time that they believed a carrier air wing would sink more than one enemy carrier on the average. It is pretty clear that U.S. aviators thought the thirty-six dive-bombers and eighteen torpedo bombers that comprised an air wing at the outset of the war would sink or put out of action (achieve a "firepower kill" on) several carriers with one cohesive strike. They estimated that the enemy could do the same. They were obsessed with the need to get at the enemy first, and we need not accept their optimism to see the enormous advantage of striking first.

The picture gets interesting when the results for *B*, the inferior force, are perused. If both sides attack together *B* cannot win, but compared with its performance in the Fiske model of continuous fire *B* does well—the enemy, while winning, can suffer severely. Even more instructive are the numbers when *B* successfully strikes first. Unlike *B* in Fiske's continuous-fire model, here *B* can be outnumbered 1:2 and establish the basis of future equality if he can attack and withdraw safely. He can be outnumbered 2:3 and establish the same after-action equality even if *A* is able to counterattack after absorbing the first blow. Evident as all this may be, to note it is crucial, since it is the basis for understanding much about the five interrelated tactical issues introduced above.

Before proceeding we should roughly calibrate attacker effectiveness by reviewing the four carrier battles of 1942 and then comparing them with the Battle of the Marianas, fought in June 1944.

For 1942 (not later) we will assume that:

— The carrier-air-wing effectiveness of every carrier on either side was equivalent.

— The defensive features of every carrier and its escorts on either side were equivalent.

— Japanese carriers physically separated should be counted. Deliberately or inadvertently they served as decoys and absorbed U.S. attention and air assets.

I indicate who attacked the enemy main force first. To compute theoretical results I show the results of all attacks, including diversionary actions, in the proper sequence. Although they do not enter into the calculations, initial and surviving carrier *aircraft* strengths are also shown.

The Coral Sea, May 1942

On 7 May the U.S. force (the *Lexington* and the *Yorktown*) sent a major strike against the little Japanese force covering the invasion force (the small carrier *Shoho*) and sank the carrier. On 8 May the U.S. force and the Japanese striking force (the *Shokaku* and the *Zuikaku*) struck simultaneously. The *Lexington* was sunk; the *Yorktown* suffered minor damage. The Japanese *Shokaku* suffered heavy damage; the *Zuikaku*, not found by U.S. aircraft, survived undamaged.

Theoretical Survivors

	After 7 May	*After 8 May*
A *Japan*	2	0
B *United States*	2	0

Battle Synopsis

	INITIAL FORCES		ACTUAL SURVIVORS	
	CV	*Aircraft*	*CV*	*Aircraft*
A *Japan*	2½	146	1	66
B *United States*	2	143	1	77

NOTES:

— The small Japanese carrier *Shoho* is counted as one-half.
— The *Yorktown*, though damaged, is counted as surviving. She fought at Midway.
— The *Shokaku* suffered heavy damage and is not counted as a survivor.
— This battle was marred tactically by very poor scouting on both sides.

Midway, June 1942

The U.S. force (the *Yorktown*, *Hornet*, and *Enterprise*) successfully surprised the Japanese striking force (the *Kaga*, *Akagi*, *Soryu*, and *Hiryu*) on 4 June. Most of the circumstances are well known, but many have not noted that the island of Midway served in effect as a highly significant decoy. After the successful U.S. surprise attack, the Japanese counterattacked, and then the surviving U.S. force reattacked.

Theoretical Survivors

	After U.S. attack	*After Japanese counterattack*	*After U.S. reattack*
A *Japan*	1	1	0
B *United States*	3	2	2

Battle Synopsis

	INITIAL FORCES		ACTUAL SURVIVORS	
	CV	*Aircraft*	*CV*	*Aircraft*
A *Japan*	4	272	0	0
B *United States*	3	233	2	126

The Eastern Solomons, August 1942

On 24 August the U.S. force (the *Enterprise* and the *Saratoga*) attacked the small carrier *Ryujo* with its three escorts, which were exposed in front of the Japanese striking force. The *Ryujo* was sunk. The Americans having taken the bait, the Japanese striking force (the *Shokaku* and the *Zuikaku*) surprised the U.S. force. The Japanese striking force was never pinpointed by the United States for a counterattack.

Theoretical Survivors

	After U.S. attack	*After Japanese attack*
A *Japan*	2	2
B *United States*	2	0

Battle Synopsis

	INITIAL FORCES		ACTUAL SURVIVORS	
	CV	*Aircraft*	*CV*	*Aircraft*
A *Japan*	2½	168	2	107
B *United States*	2	174	1	157

NOTES:

— The *Ryujo* with her thirty-seven aircraft is counted as one-half.
— The *Enterprise* was heavily damaged and is not counted as a survivor.
— Though surprised and unable to find and counterattack the Japanese, the United States had fifty-three fighters in the air, warned by air-search radar.
— U.S. aircraft losses were light because the *Enterprise*'s aircraft were able to land at Henderson Field, Guadalcanal.

— The ascendancy of the attacker is starting to wane. Although the survival of the U.S. carriers under surprise attack can be explained by many details of battle training and leadership, a trend has emerged that reflects the increasing capabilities of U.S. defenses.*

The Santa Cruz Islands, October 1942

On 26 October the U.S. force (the *Hornet* and the restored *Enterprise*) and the Japanese striking force (the *Shokaku, Zuikaku,* and small *Zuiho*) struck each other simultaneously. The small *Junyo* (fifty-five aircraft), although detached in a support unit covering reinforcements for Guadalcanal, was also able to attack the American carriers. The *Hornet* was sunk and the *Shokaku* and the *Zuiho* were heavily damaged.

Thereotical Survivors

	After 26 Oct
A *Japan*	1
B *United States*	0

Battle Synopsis

	INITIAL FORCES		ACTUAL SURVIVORS	
	CV	*Aircraft*	*CV*	*Aircraft*
A *Japan*	3	212	1½	112
B *United States*	2	171	1	97

NOTES:

— Two small Japanese carriers are counted as one-half each.

— The *Enterprise* suffered three bomb hits but was able to recover the *Hornet*'s and her own aircraft. She is counted as a survivor.

*Polmar (p. 253) goes so far as to call the battle a U.S. victory.

— The continuing terrible aircraft attrition—one hundred and seventy-four aircraft respectively for the Japanese and U.S. air wings—and the greater than theoretical survival rate for carriers indicate strengthened defenses. This is the battle in which the *South Dakota* is credited with twenty-six aircraft kills.

After the Battle of the Santa Cruz Islands both sides were reduced to a single operational carrier. The air wings on both sides had suffered grievously. In 1943 both sides husbanded their new and repaired carriers while the Solomons campaign continued to rage. The Japanese, however, were too quick in employing their naval aircraft from airfields in the Solomons and Rabaul, and they suffered accordingly. Unavoidable as the Japanese commitment probably was, the loss of naval aviators established the basis for the air disaster that overtook them in 1944.

Meanwhile, the U.S. carrier navy sorted out its air tactics, added AAW ships and AAW weapons, and built up its fast carrier task force to fifteen carriers (more than double the number at the war's onset). The Japanese succeeded only in building their carrier force back to nine (they had had ten carriers in January 1942). Qualitatively the Japanese were even more outmatched.

The Philippine Sea, June 1944

On 19 June Admiral Jisaburo Ozawa's fleet, built around all nine of Japan's carriers, attacked the U.S. fleet of fifteen carriers from four hundred nautical miles away. His plan was for Japanese aircraft to attack from beyond the range of U.S. aircraft and then continue on and land at Guam. Admiral Spruance could not simultaneously stay close to Saipan, where he was supporting the amphibious assault, and reach the Japanese. He chose to stay near the beach and concede the initiative of first attack to the Japanese. Their air attack was crushed. That evening Spruance allowed Marc Mitscher to attack at very long range—nearly three hundred nautical miles—with 216 aircraft. Meanwhile U.S. submarines sank two large carriers. The U.S. air attack, in part because of the long range, succeeded only in sinking the small *Hiyo* and heavily damaging the *Zuikaku*. In this attack the United States suffered the only significant aircraft losses of the battle; they were mostly operational, occurring on the long return flight at night.

Theoretical Survivors

	After Japanese strikes	*After U.S. sub attack*	*After U.S. counterattacks*
A *Japan*	9	7	4
B *United States*	6	6	6

Battle Synopsis

	Initial Forces		Actual Survivors	
	CV	*Aircraft*	*CV*	*Aircraft*
A *Japan*	9	450	5	34
B *United States*	15	704	15	575

NOTES:

— Mitscher's evening attack with 216 aircraft is the equivalent of three carrier deckloads. According to our rule of thumb, they should have sunk or incapacitated three carriers. In fact they knocked out two.

— Not all the carriers were large carriers. But putting the carriers in tables without distinguishing them does no violence to the comparisons and displays more familiar numbers.

Of the 129 U.S. aircraft lost, Mitscher's late-evening counterattack accounts for 100.

— In addition to overwhelming Japanese carrier aircraft losses, there were a few losses among scout planes aboard Japanese battleships and cruisers, and many more losses among Japanese planes based at Guam.

The Battle of the Philippine Sea was no longer a battle of scouting and attack. The defense had overtaken the offense. Years later, in a rare public statement, Spruance said he would have preferred to move away from the beach and attack, but his mission was to defend the beachhead. Through either wisdom or inadvertence, his defensive tactics worked. Ozawa's plan hinged on combining sea- and land-based air. By waiting near Saipan for the Japanese fleet, Spruance was able to destroy the land-based air threat, achieving numerical superiority for the carrier battle. Ozawa's shuttle tactics were foredoomed, because U.S. fighters pounced on most of the Japanese aircraft at or en route to Guam. By staying nearby, Spruance continued to totally dominate the Mariana Islands' airfields. By waiting for the Japanese, the U.S. fleet could devote all fighters to CAP. And *two-thirds* of Spruance's aircraft, 470 of them in fifteen carriers, were fighters. The U.S. had more fighters than the Japanese had carrier aircraft.

The slaughter of Japanese aircraft was due to a combination of U.S. defensive strength and the inferiority of Japan's pilots. How much the result should be attributed to one factor or the other matters little. Either was sufficient to guarantee an American victory as long as the U.S. Fleet maintained effective tactical concentration, which had become so much a Spruance trademark that even Ozawa expected it.

Had the year been 1942, Spruance should have gone after the Japanese fleet. Then offense dominated defense, and the first strike could have been expected to be an effective one. Three-quarters of a U.S. air wing comprised attack aircraft. But in 1944 circumstances were different. The concentrated U.S. carrier battle fleet had great potential to defend itself effectively. To strengthen its defense, the proportion of fighters in the air wings had been increased from twenty-five to sixty-five percent. The Japanese carriers decks, however, were still heavy with attack aircraft: two-thirds were dive-bombers and torpedo bombers. We must surmise that until the Battle of the Marianas the Imperial Navy clung to the misplaced hope of surprise and a forlorn faith in the offensive. After that battle and until the end of the war Japanese carriers were impotent, used only as decoys at the Battle for Leyte Gulf.

Many more fighters escorted a strike in 1944. What had been a battle to sink carriers in 1942 had become a battle to destroy aircraft. From June 1944 on, battles were between Japanese land-based aircraft and American carrier aircraft.

There is not a thoroughgoing appreciation of this shift of emphasis among naval critics. Certainly on the written record even Spruance grasped the shift only instinctively and Halsey saw it not at all. And Nimitz, disappointed at the failure of Spruance to sink ships because his mission was to guard the beach, revised priorities at Leyte Gulf and made the destruction of the Japanese fleet Halsey's primary mission. For his part, Halsey was overly eager to be thrown into that briar patch. In the Battle for Leyte Gulf he ran north, after carriers that had been planted by the Japanese to draw him away from the main action around the Leyte beachhead.

Resolution of Tactical Problems

What insight do these rough and ready comparisons yield with regard to the five main tactical problems?

1. *The tactical formation.* The first problem was whether to give each carrier its own screen or to put two or more carriers inside a common screen of escorts. The Japanese used single-carrier formations at Coral Sea.* But at Midway they were forced by a scarcity of escorts to double up (in 1942 Yamamoto still believed that carriers would protect battleships, not the converse). By 1944 the Japanese doubled up because too many cruisers and destroyers had been sunk. By 1944 the question was decided for the United States by numbers of a different kind. The fifteen fast carriers were combined in groups of three and four for cohesion, control, and AAW firepower.

All of which begs the question. The wisest conclusion is probably that in 1942 single-carrier screens were best because defenses were poor, aircraft could be launched and landed more efficiently when carriers had their own screens, and attacking first was the object. Single carriers separated by even as few as ten or twenty miles might escape attack, as the carriers *Zuikaku* in the Coral Sea and *Saratoga* in the eastern Solomons did.† But by 1944 something in offensive efficiency could be given up to exploit the withering defenses of the tight AAW circle. U.S. formations enclosed three or four carriers, and the entire disposition was kept close enough so that the whole

*Willmott, p. 260

†At the Coral Sea the *Lexington* and the *Yorktown* were formed by Frank Jack Fletcher to receive the Japanese air attack of 8 May inside a single screen of twelve cruisers and destroyers. They separated during thirty-knot evasive maneuvering into separately screened task groups.

fleet could be protected by a massed CAP. The decision to put two or more carriers in one formation properly rested on the effectiveness of the defense.

2. *Dispersal or massing*? The second tactical question was whether to divide the forces to the extent that mutual support was lost or significantly weakened. In 1942 the problem lay with the Japanese commanders, who were required to cover an invasion or reinforcement in all four of the big carrier battles. Their motives were mixed, for the Japanese also sought to draw out and defeat the U.S. Fleet. Admiral Nimitz, appreciating his inferiority, was not going to risk his fleet unless forced to do so. Each of the four battles had its own peculiar circumstances, but based on Yamamoto's otherwise incomprehensible manner of spreading his forces and, in at least one specific instance (the Eastern Solomons), of baiting a trap, there seems to be only one conclusion: as a leading exponent of naval aviation Yamamoto must have believed, as American and Japanese naval aviators did, that a successful surprise attack by two big carrier air wings would destroy more than an equal number of the opposition. One carrier, it was thought, could sink two or three carriers clustered together, and therefore massing two or three units against one risked three units and gained nothing. If this was Yamamoto's rationale, then he was thrice confounded. Code breaking gave the United States too much strategic intelligence. Air search radar gave too much tactical early warning. And, from the evidence, it can be said that the destructive power of a carrier air wing was not sufficient to justify the anticipation of a two-for-one effectiveness potential.

We are left with two inferences: first, that concentration of *offensive* firepower sufficient to win at one blow is always desirable and can in principle be had with modern long-range aircraft or missiles without physically massing; and second, that the decision to mass rests on the potential to enhance defense or to coordinate a concentrated first attack. If massing fails on either count, dispersal may be better than massing. But before one jumps too readily to the conclusion that, since modern missiles now possess a many-for-one offensive punch, stealth and dispersal are appropriate, some thoughtful analysis of the scouting process is first in order. We will examine this proposition in chapter 10.

3. *Offensive vs. defensive firepower.* As for the third tactical problem, whether to optimize the formation for offense or defense, the solution in World War II may be inferred from the solutions to the first two problems. As the war progressed, the U.S. Navy strengthened carrier defenses. First, fighter numbers were increased at the expense of bomber numbers. Second, AAW batteries were steadily added, the *Atlanta*-class AAW cruisers entered service, and, starting with the Battle of the Eastern Solomons, fast battleships were integrated into carrier screens. Third, the damage control of warships was emphasized and improved. Thus defensive considerations came to dominate and the destruction of aircraft became, all too subtly, more significant than the destruction of carriers.

4. *Daytime vs. nighttime tactics.* The fourth tactical problem was the dominance of the gunship at night. As early as the Battle of the Coral Sea the Japanese tried a night air attack. But it was surface night action that they continually sought. In three of the four 1942 carrier battles, the Japanese detached gunships to find U.S. carriers. Owing to American prudence or luck, the Japanese never succeeded in forcing an engagement. Judging from the 1942 night battles in the Solomons, it is well that they failed. Later, when the U.S. Navy went over to the offensive, it set up an ingenious task organization that permitted fast battleships to be detached from their carrier screening role and to form a battle line for surface action. That guns still dominated after dark is clearly seen in the climactic action in the Battle for Leyte Gulf. That huge battle is best thought of as a final and desperate Japanese effort to bring gunships within range of their targets. The last and very effective line of American defense was surface warship guns.

5. *Dual objectives.* Fifth and last, there was the sticky tactical problem of split objectives for the attacker. In 1944, when the U.S. fleet swept across the Pacific from Pearl Harbor to the Philippines in just twelve months, it was so strong that it could accompany the landing force and dare the Japanese to come out. It had a 2:1 numerical advantage in carriers, decisive in itself, and an advantage greater still when the quality of pilots and screening ships is factored in. Moreover, it was no longer necessary to attack first. Mass and unity of action were the keys to effective application of force. Battle

victory was not the issue. The issue was simply how to accomplish the objective with minimum losses and in minimum time.

In 1942 the Japanese tactical problem was not so simple. Strategic imperatives drove Yamamoto's tactics. Why were the Japanese caught at the Battle of the Coral Sea with a striking force of merely two carriers? It was because Yamamoto was in a hurry. His carriers were all busy. As figure 4-1 shows, in just four months he had spread Japanese outposts like tentacles of an octopus out to the south, where his oil supply lay; to the southwest, so he could seize Singapore and safeguard the East Indies to the west; and to the southeast, so he could seize Rabaul and safeguard the Indies to the east. He had

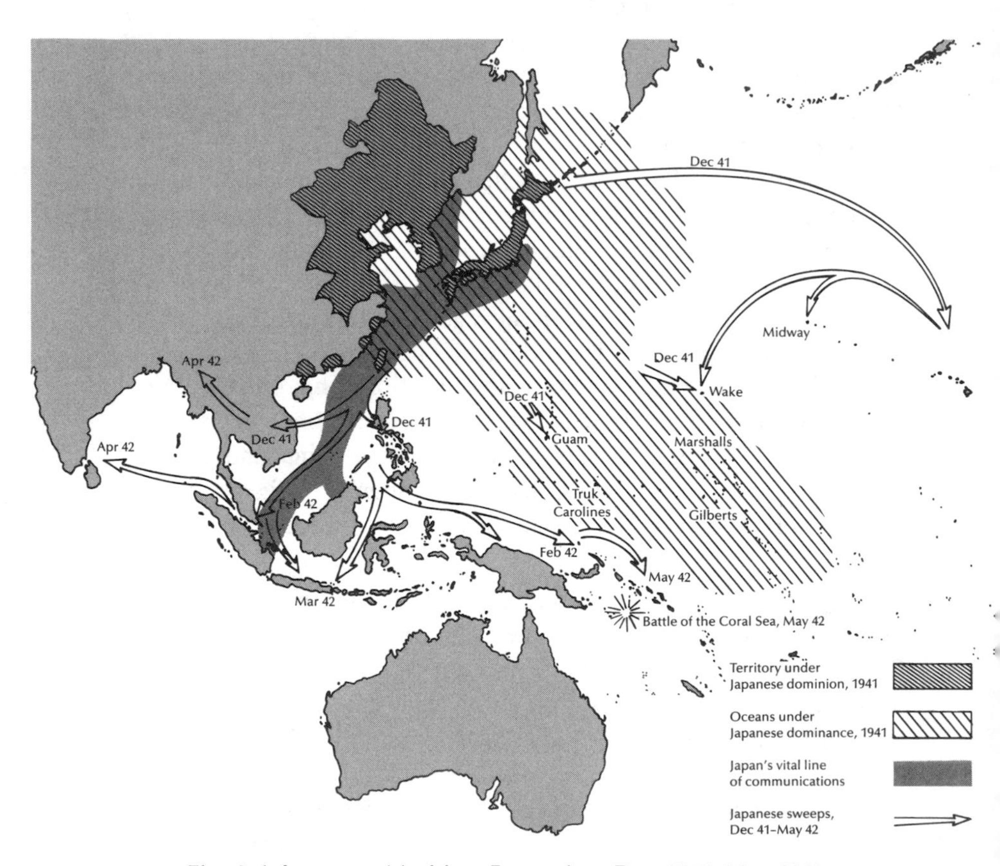

Fig. 4-1. Japanese Maritime Expansion, Dec 1941–May 1942

already secured his line of communications to the East Indies by taking the Philippines and Guam in December and January. He had eliminated Wake Island as a threat and he wanted Midway. Unlike the United States, he always envisioned supplementing sea-based air with land-based air.

Yamamoto proceeded unchecked until the Battle of the Coral Sea. Then he paid a price, though a modest one, for overconfidence. His alternative was to concentrate the Imperial Fleet's entire carrier striking power of ten carriers, which would have resulted in a glacial move southward that would have been inexorable but too slow. After all, the United States had seven big carriers flying as many aircraft as his, and soon they would be coming.

Still, while Yamamoto consolidated his network of air bases, he wanted—nay, needed—to entice the U.S. fleet into battle. In 1942 he could only do that with an invasion threat, the same way the United States planned to draw out the Japanese fleet in 1944. The Battle of the Coral Sea illustrates his priorities. The April operation was intended for the establishment of outposts at Tulagi and Port Moresby to screen Rabaul and threaten the U.S. link to Australia. Yet when the two U.S. carriers appeared and fought back, Yamamoto blasted Vice Admiral Shigeyoshi Inouye, who was in tactical command, not for pulling back the invasion force headed for Port Moresby but for pulling back from pursuing the surviving carrier. He ordered the *Zuikaku* to go plunging fruitlessly in pursuit of Fletcher in the *Yorktown*. At this stage of the war American naval leaders were no wiser than Yamamoto. The U.S. Navy had two carriers tied up in the Doolittle Raid against Tokyo, which is why the *Lexington* and the *Yorktown* were all that was available (the *Saratoga* had been crippled by torpedoes from a submarine).

Yamamoto split his carriers for the Midway operation because he was in a hurry and his motives were mixed. With two geographical objectives, Midway and Kiska/Attu, he had to cover two invasion forces. But the Kiska/Attu force was being used as a diversion, the Midway invasion force to draw out the U.S. Fleet. Yamamoto wanted the carrier striking force ahead of the invasion force. That was well and good, but he still thought of the battleships as final arbiters, there to mop up and too precious to expose until command of the air space was settled. He need not and should not have squandered effort in the Aleutian sideshow, which drew away two small carriers. It was right, however, not to wait for the repair of his damaged carriers. By all reasonable estimates there could have been at most two carriers plus Midway's aircraft facing him. If he had waited thirty days more, the *Yorktown*'s wounds would have been healed, the *Saratoga* would have joined up, and in another thirty days the *Wasp* would have arrived from the Atlantic.

History has been too eager to second-guess Yamamoto's decisions. He lost at Midway for all the following reasons:

— The U.S. Navy had strategic intelligence.
— Vice Admiral Chuichi Nagumo had no air search radar.
— Japanese scouting was mediocre. Nagumo launched a feeble air reconnaissance and the Japanese were beyond help from land-based air reconnaissance, a rare thing for them. And Yamamoto's subs arrived too late on picket stations.
— Spruance was skillful tactically.
— The American air wings were brave.
— The American air wings were lucky.

Take away any of these six conditions and it is more than likely the Japanese would have destroyed the American fleet and captured Midway. If good luck influences battle results, then results are not a perfect indicator of planning soundness. Historians must not evaluate the tactician's acumen solely on the outcome of the battle.

Still, the Battle of the Coral Sea was Yamamoto's clue that it was time to proceed more carefully—to make sure his operational plan was designed to achieve its objective. At the Battle of Midway the foremost objective was, or should have been, to draw the American fleet into a battle it would have to come out and fight. To split off two small carriers for the Aleutians operation was a mistake. The strategy to extend the Japanese defensive perimeter had stretched tactical capabilities too thin. It was time for tactics to counsel strategy, and tactics said that the Japanese navy had too many objectives.

Though we are not likely to see the Pacific war over island air bases reproduced, we can anticipate the recurring tactical problem a commander with superiority faces against an enemy who knows his inferiority and declines battle. When an attack on a land target is the way of drawing out an inferior enemy, it is too easy during planning to let the land attack become the end itself and to forget that the attack is but the means to a greater end, in this instance, of destroying the enemy's seagoing forces.

Summary

In World War II aircraft became the chief naval weapon during daylight hours because of their effective range and their (not limitless) integral capacity for scouting, guidance to their target, and coordination. Not just any type of aircraft could do the job, and considering the combat loss rates, not just any pilots.

The polarized overestimates and underestimates of air strike effectiveness against warships that were made before the war led to unexpected tactical results during the war. Combat leaders, who by then had hard evidence that aircraft were effective against ships in daylight, had to learn the limits of the effectiveness—which we have deduced was never more than about one carrier sunk per air wing attack—and change their tactics accordingly.

Meanwhile the balance that favored air offense at the outset shifted back in favor of ship defense as the war progressed, and that also had to be appreciated for its tactical significance. It boiled down to a central tactical issue. If one carrier could sink two or three in an effective attack against any defense, there was no point in massing two or three carriers, which might be located together and sunk. If, however, it took two or more carriers (or repeated carrier strikes) to sink one, the *concentration* of enough striking force was essential (but it was only necessary to *mass* force if separate carrier task forces could not be coordinated by means of C^2). As the war progressed, a third case came to dominate in which the defensive firepower of two or more carriers operating in proximity could be massed to great effect. Concentration of force for offensive action became an automatic side benefit and the problem of coordinating separated forces disappeared. It was defensive considerations that drove the decision to mass and eliminated the need for stealth, deception, and divided forces.

The Japanese had to attack effectively first. A simultaneous exchange of attacks with similar losses to both sides would ruin them in the long run because they could not afford to exchange carriers one for one. They had to attempt stealth, deception, and (probably) divided forces as a calculated risk. They gambled, likely even believed, that one carrier could sink two. Even though wrong, this was a good gamble at the beginning of 1942. As early as late 1942 it was a very bad gamble.

We have discussed many reasons for the resurgence of defense. AAW guns alone could have been enough to cause it, but the final and decisive factors responsible for the success of American defense were things the Japanese could not possibly fold into their early planning, namely, radar and cryptanalysis. Except in the Battle of Britain, nowhere was radar more quickly put to decisive use than in the Pacific carrier battles. Cryptanalysis for its part almost eliminated the chance of the Japanese achieving surprise. Stealth and deception were foredoomed. Under the circumstances the Japanese might as well have massed and taken their chances, especially in 1942 when they had numerical superiority and qualitative equality. By 1944 nothing they could do mattered. The Japanese would have less and less to show for their efforts, whether they were expended against a conservative enemy like Spruance off the Marianas or a rash enemy like Halsey at Leyte Gulf. U.S. defensive strength ensured that the American fleet would survive long enough to counterattack, and U.S. offensive advantage ensured that the Japanese would perish.

5

World War II: The Sensory Revolution

Scouting Measures and Countermeasures

Many of the postmortems of World War II express more awe of the sensors that revolutionized naval combat than of the weapons, including the airplane, that fought it. But any treatment of sensors, weapons, and tactics in isolation is artificial; all three together decided battles. To get at the way the new sensors affected tactics and weapon performance during the war, the first step is to establish a framework for discussion. Then we are able to look at radar as foremost among the new scouting devices. Night surface combat of 1942 and 1943 in the Solomons campaign will serve as a premier illustration of radar's tactical potential and its underexploitation by the American cruisers and destroyers that fought those action-packed battles. Nearly as important was the rise of communications intelligence in World War II. A brief look at the submarine war in the Atlantic, with emphasis on sensors, will show the great extent to which what might be called the sensor war was the decisive element in the defeat of the German U-boat. The full significance of cryptology as an element of scouting measures and countermeasures can be better appreciated now that so many closely held accounts of code breaking in World War II have at last been declassified.

The new sensor war was a duel. As the reach of weapons increased, electromagnetic science raced to stay abreast, inventing the means of detecting and communicating at long range. Concurrently, countermeasures to thwart the new electromagnetic technology were being worked on. It is useful to think of the signals put out by these electromagnetic systems as measures that contributed to force effectiveness in either of two ways: aiding in the detection, tracking, and targeting of the enemy (that is, scouting) or in the execution of a commander's battle plans (that is, controlling). Simultaneously, the force was trying to diminish the effectiveness of the enemy's scouting and control systems.

One common way to categorize these countermeasures is to divide them into techniques for destroying, disrupting, deceiving, denying, or exploiting enemy signals. The possible actions, by no means exhaustive, may be laid out as in table 5-1.

Table 5-1 is a static display. It gives no sense of pace or timing of signals exploitation, of buying time or intruding in the timing and coordination of enemy action. Except subconsciously, military men do not think in the destroy, disrupt, deceive, deny, exploit framework but leap at once to the possibilities of manipulating their own and the enemy's specific equipment and command structure. A more useful way to visualize signals exploitation is first to lay out the sequence of measures that must be taken for a fully effective attack, as table 5-2 does.

Evidently some scouting systems can perform more than one step. A surveillance-communications satellite could perform measures *A*, *B1*, *B2*, and *B3*. One of the advantages of aircraft has been that they can perform the tracking, targeting, attacking, and damage assessment functions in a single flight. By way of contrast, external scouting for surface-to-surface missiles must perform functions *A*, *B1*, and *B2* before an attack. To ensure full effectiveness, thorough scouting may be necessary. To be sure of hitting his enemy's carrier or carriers, the attacker may need to know with precision not only the enemy's position but also his formation. In addition he will need to know whether his enemy has cruise missiles of his own, located on ships other than carriers, that can be used in a punishing counterattack.

In principle, the chain of measures necessary for an attack can be broken anywhere by one countermeasure. A successful countermeasure can defeat, delay, or reduce the effectiveness of the attack by denying the enemy any step in the sequence. It is not our purpose here to pursue the measures and countermeasures in detail. Suffice it to say that each side strives to maintain the chain from *A* to *C* with some combination of redundancy, cover, covertness, cryptology, and

Table 5-1. Countermeasures Against Enemy Signal

	Enemy Signal Function	
	Scouting (e.g., radar)	*Controlling (e.g., radio)*
Destroy	Attack it	Attack it*
Disrupt	Jam it	Jam it†
Deceive	Provide false targets	Provide false targets‡
Deny	Avoid the sensor§	Use covert transmission‖
Exploit	Counterdetect enemy	Monitor enemy's traffic#

*In ships at sea, attacking the weapon system also usually meant attacking the signal system. From this derives the current interest in offboard command, satellites, land-based over-the-horizon radar, and other approaches, all of which will complicate tactics.

†Jamming a scouting system usually buys range. Jamming a controlling system usually buys time. Either translates into a sought-after positional advantage.

‡Deception also can be used against enemy *weapons*. Chaff used against missiles and noisemakers against homing torpedoes are prominent examples.

§Submarines are immune to visual, radar, and infrared detection and sometimes, through quietness, to passive sonar detection.

‖There is asymmetry here, because covert signals like low-probability-of-intercept (LPI) radio transmissions are *measures* to confound major enemy countermeasures, namely, intercept and exploitation. Since this table's utility is limited, I have not pursued the logic of the problem further.

#Exploitation may be the most important component of this table, considering the serious possibility that the enemy will detect a scouting system before the scouting system detects the enemy. And radio transmissions may be heard and decrypted. In addition, the threat of counter-countermeasures enters in when the enemy suspects his traffic is being listened to. In that case, he can plant misinformation, taking due care to avoid self-deception. While there is a long history of strategic deception with assorted agents and double agents of espionage to do the communicating, this is not our concern. For an excellent and rigorous study of this subject, see Daniel and Herbig's *Strategic Military Deception*.

sheer electromagnetic power while attempting to break the other's chain decisively at its weakest link. The sensor war aims to shorten the time it takes to complete one's own chain, and it aims to lengthen the enemy's time. A modified table of measures, 5-3, emphasizes the relative consequences of a successful countermeasure.

Observe that the countermeasure called exploitation has no place in table 5-3. That is because its effect is different from that of destruction, disruption, deception, or denial. Exploitation enhances one or more of the measures *B1*, *B2*, *B3*, and *B4*, which permits one's own attack. Exploitation has its own dynamics, and fits into the

Table 5-2. Measures Required for Effective Attack

Strategic detection (A)	— observation of major enemy force presence or future presence in a region
Tactical detection (B1)	— location of the enemy for the purpose of attacking him
Tracking (B2)	— knowledge of enemy position sufficient to launch a successful attack
Targeting (B3)	— determination of enemy dispositions in detail sufficient to attack with maximum effectiveness
Attacking (B4)	— control of a coordinated, concentrated attack
Damage assessment (C)	— post-attack evaluation of the results

Table 5-3. Effect of Successful Countermeasures

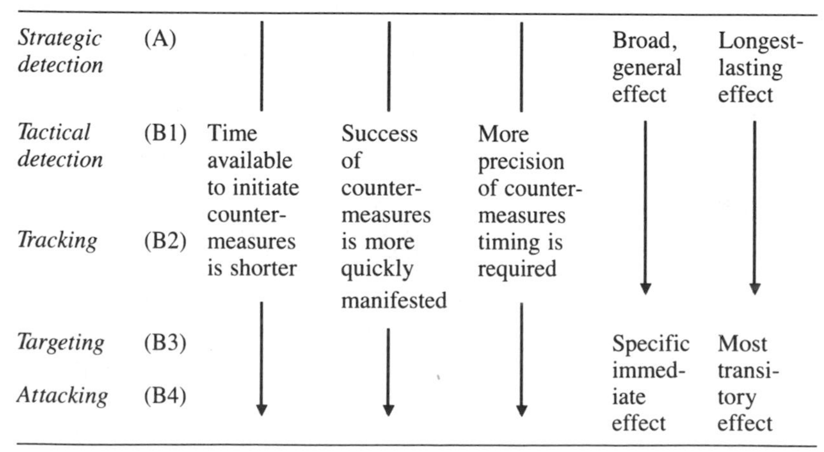

Strategic detection	(A)	↓	↓	↓	Broad, general effect ↓	Longest-lasting effect ↓
Tactical detection	(B1)	Time available to initiate counter-measures is shorter	Success of counter-measures is more quickly manifested	More precision of counter-measures timing is required		
Tracking	(B2)					
Targeting	(B3)	↓	↓	↓	Specific immed-iate effect	Most transi-tory effect
Attacking	(B4)					

tactical picture in a way best shown in the force-on-force model of modern warfare described in chapter 10.

For the purpose of seeing the several roles played by the new sensors in World War II and the signals warfare they fostered, the list of measures required for attack is sufficient. Playing the roles are cryptanalysis, used in a startlingly effective way for strategic positioning; search radar for active detection and tracking; passive signals interception, especially for tactical detection but also for other functions; sonar for detection, tracking, and targeting; and rapid, disciplined voice communications for attack. Let us observe these dynamic processes and how they worked as signals warfare shot into prominence in World War II.

Radar

Radar and radar countermeasures were the most important of the sensory tools of warfare that came of age in World War II. In our discussion we should also include the proximity fuze (which was a tiny, shock-resistant radar) and its enhancement of weapon effectiveness. Against aircraft that were closing on a five-inch dual-purpose gun, the proximity fuze not only increased the allowable fire-control dispersion error by two or three times but also simplified a three-dimensional fire-control problem, making it two-dimensional.

The possibility of using pulsed radio waves to detect ships and aircraft occurred to scientists well before World War II. Several countries had secret research under way in the 1930s. In December 1935 the first five radar stations were established by Britain on the east coast of England. Though radar was indispensable in the aerial Battle of Britain, it was not a secret weapon in the sense that cryptanalysis was, that is, a wartime tool unsuspected by the enemy. The presence and significance of radar was quickly appreciated by all sides. Radar and radar countermeasures were part of a fast-paced technological race drawing on vast scientific resources, including, for example, a one-hundred-fold buildup of personnel at MIT's radiation laboratory.*

*The number increased from forty to four thousand (Brodie and Brodie, p. 209).

As a tool of war, radar was ubiquitous. By the end of 1939 shipboard prototypes were being tested for long-range aircraft detection, antiaircraft fire control, and surface tracking. A series of remarkable breakthroughs came with the British and American collaboration which began in 1940. Centimeter-wavelength radars were ready for production in 1942, and they had sufficient definition to be used for detection of single aircraft, for fighter direction, day or night, and for accurate surface and AAW gunlaying. By 1943 radar was fitted in enough reconnaissance aircraft to have a major influence on search and attack against surface ships, and in enough antisubmarine patrol aircraft to reverse the momentum of the U-boat campaign in the Atlantic. From 1940 on radar was vital to fighter defenses over land, and during the offensive fighter and bomber sweeps before the Normandy invasion it was the key to their effectiveness. For antiaircraft defenses radar was just as important over land as at sea.

Radar quickly became an indispensable navigating tool. It allowed high-speed surface operations in narrow seas, and it came to be so much relied on that when a ship lost her radar at night she was literally and psychologically lost. As a more direct instrument of war, radar started helping guide bombers to their targets over Germany in 1943. Loran was used to aid ship and aircraft navigation everywhere operations were intense.

Radar was developed in the United States, Britain, Germany, France, and Japan. As aircraft offensive and defensive operations became dependent on radar, measures to counter it acquired the highest tactical significance. The first great sensor war occurred in the air over Europe. World War II—specifically, from the time of the Battle of Britain to 1945—offers the best case study of the measures, countermeasures, and counter-countermeasures taken in scouting and weapon delivery (it is better for study than the Vietnam War because analysts have been denied North Vietnam's tactical picture).

We have already seen the big edge radar gave the United States in the air over the Pacific. To demonstrate what radar meant, at the Battle of the Eastern Solomons, U.S. air search detected the approaching Japanese at eighty-eight miles, far enough out that fifty-three fighters—the works—were scrambled with full fuel tanks. Radar gave the United States time to put interceptors in the air after the raid was detected; all fighters were vectored to the attack aircraft without fear of surprise from another quarter.

Because the Japanese were slow to develop radar, the U.S. Fleet had an unparalleled opportunity to exploit its surface search advantage well into 1943. Even after that, until the end of the war, it was technologically well ahead of the Japanese fleet. In the Solomons, long-range patrol aircraft and coastwatchers gave the United States early warning of the approach of almost all Japanese warships. (Cryptanalysis seems to have played an insignificant role in this strategic detection since Japanese naval codes had just been changed.) Because Japanese surface warships had to be well clear of U.S. air cover by daylight, the United States knew within a couple of hours when they would arrive. In eleven major engagements from August 1942 to November 1943, radar gave the United States the means to detect, track, and target an approaching surface force and blast it out of the water with guns and torpedoes before it knew of the American presence. There was, however, a clumsy lack of recognition on the part of the U.S. Navy that radar offered unparalleled opportunities and required new tactics. And there was a second problem that American tactical commanders had to solve at the same time. The Japanese had a secret weapon too. It was the Long Lance torpedo.

Night Surface Actions in the Solomons

The tactics of the night battles were a competition between these two new tools of war, each capable of exploitation: American radar and the highly lethal long-range torpedoes of the Japanese. In the five battles from August to November 1942 Japanese preparations paid off. The Imperial Navy developed a coherent system of night tactics well before the war and practiced it assiduously. Night action was part of Japan's prewar recipe of equalizers designed to whittle down the U.S. fleet before a conclusive battle line engagement. U.S. Navy practice before the war concentrated on daylight fleet engagements built on the column of capital ships. The American tactical concept was to seek a position that would facilitate (or at least not hinder) capping the enemy T. American tactical training proved counterproductive in the Solomons.

In a strategic context, the Solomons campaign from beginning to end was a contest over airfields on land and flight decks at sea, and for control of the air space around them. For the first six months, from August 1942 to January 1943, the campaign centered on Guadalcanal. In daylight the United States controlled the air around Henderson Field. The Japanese had the same advantage around Rabaul, New Britain, six hundred miles to the northwest. But when the sun set air power lost its grip, and surface combatants met and fought once more. Every night the Japanese threatened to rush warships south through the narrow waters of the Slot between the dual island chains that constituted the Solomons. To the Americans the enemy warships were the fearsome Tokyo Express, which was bent on delivering reinforcements to Guadalcanal or a brutal gunnery bombardment to Henderson Field. Whenever American warships tried to stop the Japanese the result was a lethal surface battle in the blackness of night.

It is well to pause long enough to stress that in naval duels one side or the other nearly always has the complicating problem of a beachhead or a convoy to protect. Every night battle in the Solomons was fought over some objective on land. At the first, Savo Island, the United States was defending the beachhead at Guadalcanal. At the next to last, Empress Augusta Bay, it was defending Bougainville. In the battles between those two the Japanese had the enduring problem of fighting an engagement while trying to reinforce or withdraw a garrison in the Solomons. Two carrier battles, the Eastern Solomons and the Santa Cruz Islands, were tied completely to events on Guadalcanal. The flawed strategy of piecemeal commitment of force in the Solomons gave Japanese tactical commanders the harder task. Keeping reinforcements flowing to sites where ships could not stay was hard, because Japanese warships could only move for a few hours around midnight. It was a sharp disadvantage that American tacticians failed for too long to exploit.

The Japanese did well in the early battles (August 1942 to July 1943) despite their handicaps. This was because:

— The United States failed to grasp that the killing weapon was the torpedo.
— The United States had no tactics suitable for night battle at close quarters.

— The United States was slow to learn. Because of the rapid turnover of tactical leaders, the *pace* of the battles overwhelmed the Americans.

— Above all, the United States did not exploit its potentially decisive radar advantage—the edge in first detection and tracking that surface-search radar gave and in targeting that fire-control radar gave. While not all ships had both advantages from the start, the radar equipment there was should have been better utilized.

From August 1942 to July 1943 the U.S. Navy suffered from these four shortcomings. From August 1943 to the end of the surface fighting in November 1943 it finally took advantage of the latent potential of radar, using new and compatible tactics.

Phase One, August 1942–July 1943

From the outset Japanese tactics were usually to approach in short, multiple columns, get all ships into action at once, and maneuver in defense against torpedoes. Sometimes destroyers would be positioned ahead as pickets to avoid ambush. On detecting an enemy force, they would close, pivot, fire torpedoes, and turn away. Sometimes they would not fire their guns at all.

The U.S. tactic was to use a long, single, tightly spaced column. The navy expected and achieved first detection and tried to position its column so that all guns would bear across the enemy's axis of approach, crossing his T. A range of ten thousand yards would, it was thought, be safe from torpedoes and perfect for guns, and if the enemy held to a steady column, the battle would be settled by guns before torpedoes entered the picture (if they did, it would be at a conjectured effective range of under five thousand yards).

But the range closed too fast, the Japanese would not stand still, and their torpedoes were devastating. At first U.S. tactical commanders did not put themselves in radar-equipped flagships, commands were vague or tardy, and the battles were fought at point-blank range, sometimes in great disorder. Later the Americans learned to respect the deadly effectiveness of a massive torpedo barrage against a long, tightly spaced column, but because the actions were closer than they wanted or expected, they didn't grasp the fact that the Long

Lance in barrage was effective at ranges equal to the effective range of cruiser and destroyer guns.*

The Battle of Cape Esperance, fought at the northern point of Guadalcanal on the night of 11–12 October 1942, was the first time the United States could put together a force to take on the Tokyo Express after the debacle of Savo Island five weeks earlier. It illustrates the phase-one tactics of both sides and the tempo of these actions. The United States had a nine-ship column capping the Japanese T, crossing ahead of an approaching force of three cruisers and two destroyers in "perfect" position. The four American cruisers were spaced at six hundred yards between ships, and the five destroyers were at five hundred yards. The light cruisers *Helena* and *Boise* had their SG surface-search radars on. In Admiral Norman Scott's flagship, the *San Francisco,* the radar was off because it was an older model with a longer wavelength susceptible to Japanese intercept. Let us watch the speed of events as the battle unfolds (see figure 5-1).

2325 The *Helena* detects at fourteen nautical miles. Spotting planes are in the air but ineffective. The *Helena* ponders, reports nothing. Seven minutes elapse.

2332 Scott reverses course for tactical reasons that are sound only because he is unaware of the enemy closing at nearly half a mile

*A quick example of torpedo fire: Japanese spreads were thrown in the general direction of U.S. gun flashes (sometimes search lights). A column of eight ships was about four thousand yards long. At a distance of four to eight thousand yards, the torpedo spread was more than likely to be inside the end points of the target column. Ships were over one hundred yards long and spaced at five to six hundred yards, so about one weapon in six would hit. With as many as thirty torpedoes in the water, unseen and unsuspected, the results could be, and were, lethal, especially since one hit on a cruiser or destroyer nearly always achieved a firepower kill. McKearney's study (1985) found that the average torpedo firing range over the entire campaign was eighty-five hundred yards, which made the running range on the order of seventy to seventy-five hundred yards (p. 154). He concluded that the average hit probability for all engagements was .06, but at the battles of Tassafaronga and Kula Gulf torpedo hit probabilities approached .20 (compiled from appendix A, pp. 188–246). Not all U.S. columns were as long as eight ships, not all torpedoes would have been reliable, and toward the end of the campaign the Americans learned to comb the tracks, so his numbers corroborate the theoretical estimate.

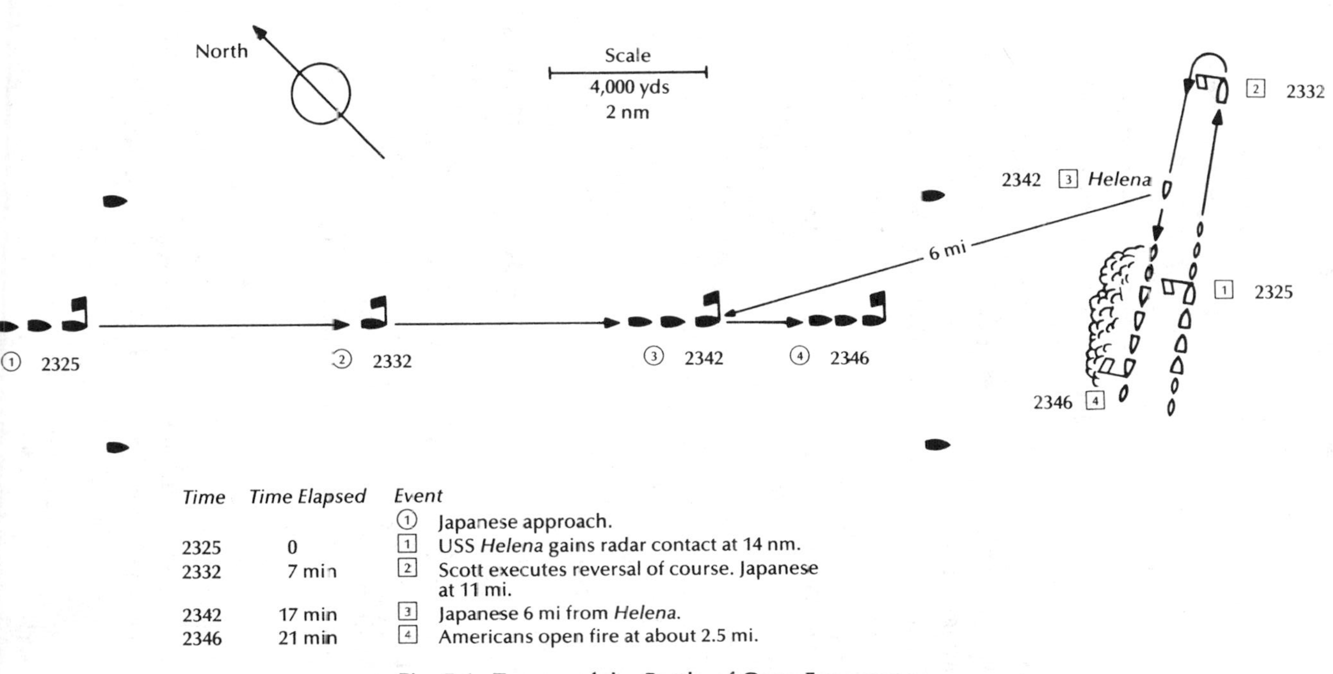

Time	Time Elapsed	Event
		① Japanese approach.
2325	0	[1] USS *Helena* gains radar contact at 14 nm.
2332	7 min	[2] Scott executes reversal of course. Japanese at 11 mi.
2342	17 min	[3] Japanese 6 mi from *Helena*.
2346	21 min	[4] Americans open fire at about 2.5 mi.

Fig. 5-1. Tempo of the Battle of Cape Esperance

a minute. But he executes a strange maneuver that puts the three lead destroyers directly on the engaged side of his cruisers, racing to regain their van position. Ten minutes elapse.

2342 The CO of the *Helena* reports the enemy at six miles directly on the starboard beam, a position ideal for gunfire except that, without radar, Scott can't see his three destroyers. Scott, in a quandary, asks the destroyer division commander where his destroyers are. Captain Robert G. Tobin says he is dead on the starboard beams of the cruisers, right in the middle. He is correct about only two of his destroyers, because the third, the *Duncan*, is not astern. Having seen the enemy on radar at four miles and believing Tobin is charging them, the *Duncan* is headed toward the enemy and now in the middle of no man's land. Three minutes elapse.

2345 The five Japanese ships, still oblivious to their danger, are at two and a half miles and visible in the *Helena*, which with radar contact knows where to look. The captain asks permission to open fire. The signal is ambiguous—Scott thinks the *Helena* is asking whether another voice radio transmission is rogered by Scott.* Scott says affirmative and is shocked to see the *Helena* open up with fifteen six-inch and four five-inch guns. One minute elapses.

2346 The Japanese are more stunned than Scott. Until that instant oblivious to the presence of U.S. ships, they are as vulnerable as U.S. ships were at Savo. They corpen-18 away and for once fail to fire torpedoes.† The range is two miles, point-blank. One minute elapses.

2347 Scott orders check fire, even as all ships but his own flagship open up. The order is not without cause, because the *Farenholt* and *Duncan* are in line of fire and will show six- and five-inch holes on their port sides after the battle. Four minutes elapse.

*Interrogatory Roger was the ambiguous signal book query.

†A corpen is a follow-the-leader maneuver. Corpen-18 means to reverse course by turning in succession 180 degrees.

2351 Scott orders resume fire, but most ships have not stopped. His ships have no fire distribution plan and concentrate on what they can see, two already-crippled destroyers ablaze. Gunfire distribution is a problem that will never be solved.

The Japanese fled, and no one seems to have suggested a hot pursuit. Four cruisers and five destroyers had totally surprised three enemy cruisers and two destroyers. The United States sank or damaged two destroyers and one cruiser and had two destroyers and one cruiser of its own sunk or crippled. Much of the damage was self-inflicted. Before the *Duncan* sank, the hits made in her by U.S. gunfire could be seen.

The U.S. Navy counted Cape Esperance a victory; with U.S. firepower potential and advantage of initiative, it should have been an annihilation. In those early days of the sensory revolution the Americans used radar, radios, electronic countermeasures (ECM), and the signal book ineptly. At Cape Esperance they did not fire a torpedo. After that the enemy was never caught so unprepared, so unready to counterattack, or so lightly loaded with the killer weapon, the torpedo.

It was a pickup force the United States had. Morison said the force had no battle plan, but in fact the gun-column was the plan. There was no sense of pace in the decisions made between first radar detection and the time for attack. What was worse, because Cape Esperance was an apparent victory, it seduced the United States into using the same tactics at Tassafaronga and later battles.

Six weeks later the Battle of Tassafaronga exposed all the U.S. shortcomings: the scratch team, inexperienced leadership, and obsolete, tightly spaced single column. Again the Americans had an overwhelming force of five cruisers and six destroyers against eight destroyers, six of which were laden with supplies, and they achieved total surprise, detecting with radar in ample time. Gunfire was opened with radar fire control at the ideal eight to ten thousand yards. But, because of an inadequate sense of timing, torpedo fire was held too long, and the course was held because the Americans supposed they were out of enemy torpedo range. This time they faced the redoubtable Rear Admiral Raizo Tanaka, who had drilled destroyer teams since 1941. His destroyers were trained when surprised to wheel and fire their Long Lance torpedoes. They did, and as a result four of the five U.S. cruisers were sunk or damaged and Tanaka lost only a picket destroyer. The U.S. commander was Carleton Wright, but the plan was not his, for he had relieved someone else two days before the battle. When it was over Nimitz said the lesson of the battle was "training, training, and TRAINING." But the tactical lessons still slipped through American fingers.

Phase Two, July–November 1943

It was a full year after Savo Island and three battles later that the U.S. Navy got it right. The Japanese continued to operate as before, except more proficiently and with better scouting: they used night air reconnaissance, radar ECM, and rudimentary radar. The United States would take on a tougher enemy but it would also fight with more skill. First, it had trained units. Second, it trained to sound tactics. Torpedoes would be fired by small, compact divisions of three or four destroyers. Two divisions would be sent in; one would fire and turn away, then the other would fire, then both would mop up with gunfire. When cruisers were present, they were to be kept at a distance of more than ten thousand yards for fear of the deadly Japanese torpedoes. But it was best to let the destroyers do the damage: they had the killing weapon.

It was at the Battle of Vella Gulf on 6–7 August 1943 that U.S. tactics came together. The navy had its usual foreknowledge of a Japanese reinforcement mission. This time destroyers alone were sent hunting, and Commander Frederick Moosbrugger had the mission, the firepower, the scouting advantage, and the tactics to fuse all three. According to E. B. Potter, Moosbrugger used Arleigh Burke's battle plan.* It has Burke's flair. The plan looks like the scheme in figure 5-2. There were two units, both with three destroyers, well drilled and tactically disposed to stay out of each other's way with the help of radar. They would maneuver with lightning precision. The first division under Moosbrugger carried forty-four torpedoes, and it would fire as many as possible. The second triplet of destroyers under Com-

*Potter, p. 313.

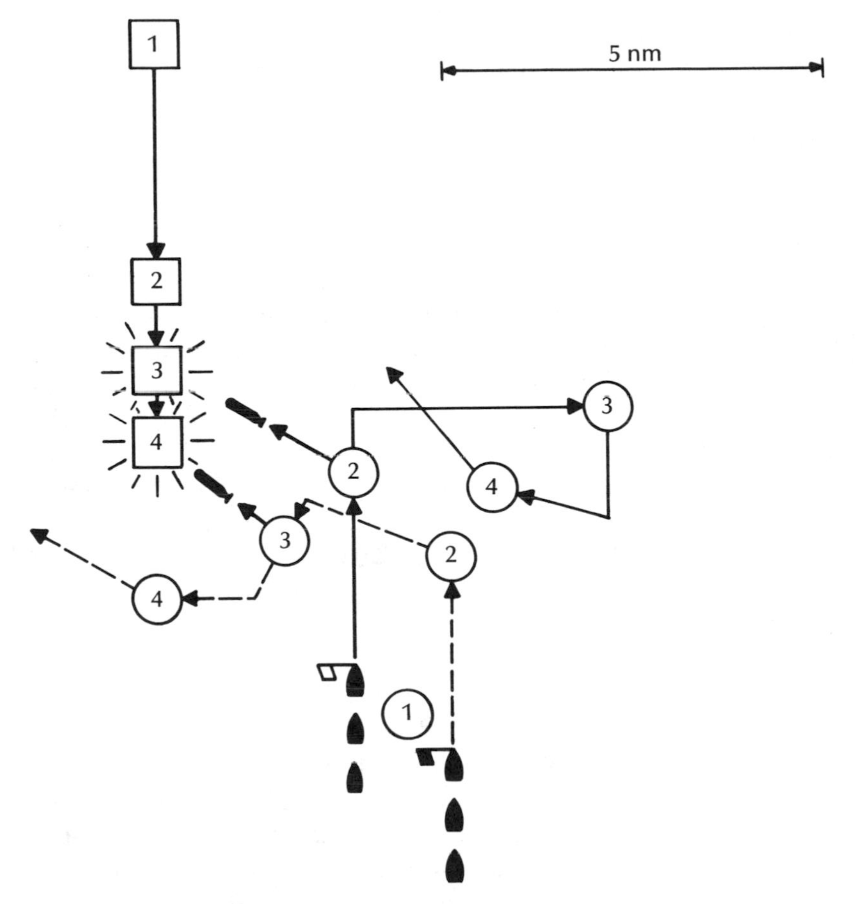

1. Detection range: 19,000 yds. Time: 0.
2. All possible torpedoes launched by first unit. Time: +6 mins.
3. Second unit closes, wheels, and launches torpedoes.
   First torpedoes impact. Time: 12 mins.
4. Second torpedo impact: Move to mop up with gunfire in prolonged stern chases. Watch out for torpedoes!

Fig. 5-2. The Battle of Vella Gulf, 6–7 Aug 1942

mander Rodger Simpson, being heavier in AAW batteries, carried only twenty-four torpedoes. Both divisions would move in, bows to the enemy torpedo threat, and wheel and launch their fish. Like Japanese tactics, all this would be done stealthily, without gunfire. After the shock of the torpedo barrage was over, gunfire and aggressiveness could be used in proportion to the damage done, but at that stage the destroyers would have to watch out for the death sting of the enemy.

On the night of 6 August tropical rain beclouded the SG radars of Moosbrugger's short twin columns. The islands around Kolombangara and Vella Lavella confused the radar returns and presented the usual complications of a military operation. Having been spotted by an American night-search plane earlier, the Japanese were on the alert for a U.S. force. Their mission was to reinforce the Kolombangara garrison. They had four destroyer transports, a 2:3 disadvantage. Those were odds they had beaten before.*

Moosbrugger's force detected the Japanese at nineteen thousand yards. Every ship was told at once. Moosbrugger shifted course right thirty degrees by twin corpens to hit the four enemy destroyers port to port. Gun and torpedo fire-control solutions were cranked in as the unseeing enemy cooperated with steady course and speed. Just seven minutes later the range was four miles. With lookouts estimating the visibility at two miles in the blackness of night, it was time to act. Moosbrugger gave the order to fire, simple and unambiguous, setting all the tactical flywheels of his semiautomatic plan churning in synchronization. The portside torpedoes—twenty-four in all—hit the water from three ships. The firing range was just sixty-three hundred yards for a running range of four thousand yards—as good a setup as could be expected and all one could ask for. A minute later a "turn nine" order came from Moosbrugger. His division executed a simultaneous ninety-degree turn to starboard to clear out, combing the wakes of the predictable enemy torpedo counterattack. At the same time Simpson wheeled his three destroyers to port and bored in.

The men in three Japanese destroyers scarcely knew what hit them.

*However, neither side had very accurate estimates before that battle or any other.

On the U.S. side there was tension aplenty to grip the Americans, who could well remember the chagrin of earlier battles lost during what should have been the mop-up phase. But not this night. As in all good naval battles, the outcome was decided by a feasible plan whose tactical cohesion came from training, good scouting, and the swift thrust of a killer weapon. Three of the four Japanese ships were sunk at the cost of an American gunloader's crushed hand. Like Nelson, Moosbrugger made sound tactics look easy.

The U.S. Navy had found the tactics to match its radar advantage and neutralize the enemy's torpedo advantage. It could now beat the Japanese with torpedoes, their own superior weapon. Good sensors and scouting could overcome better firepower. And the United States would win with little ships because they carried the big weapon. The destroyer's torpedo, not the cruiser's gun, ruled at night. Hit and move was the answer, not crossing the T; units had to be nimble rather than fixed in a sturdy, steady, cohesive—and suicidal—column.

There would be a setback at the Battle of Vella Lavella on 6–7 October 1943. That was another case of an unblooded commander with a scratch force—a nominal number of six destroyers, in fact only three, facing a nominal number of nine, in fact six, Japanese destroyers. Captain Frank Walker, the American commander, chose to go in with his three ships rather than wait for three more to join him from ten miles away. He had the usual radar advantage, fired fourteen torpedoes at seven thousand yards, then gave the game away by opening up with guns while his fish were running. Staying broadside to the enemy, his three destroyers paid the penalty: two of them were torpedoed and the third collided with one of the victims. Three destroyers were put out of action in exchange for one Japanese destroyer sunk—the one that turned into the U.S. torpedoes instead of away.

But then came the masterful battles of Empress Augusta Bay and Cape St. George, fought by tacticians par excellence Stanton "Tip" Merrill and Arleigh Burke. It is not necessary to recount all the details. Empress Augusta Bay (2 November 1943) looks less decisive in the box score than it was in fact: a Japanese light cruiser and a destroyer were sunk and a heavy cruiser was damaged in collision in exchange for one U.S. destroyer damaged. Merrill's mission was to defend the landing beach on Bougainville. His force of four light cruisers and eight destroyers was pitted against an enemy that we now know consisted of two heavy cruisers and eight light cruisers and destroyers, equal forces on paper. The tactical plan was for Merrill to keep his cruisers at longer range, about sixteen thousand yards, executing 180-degree turns timed to upset Japanese torpedo-fire-control solutions, while staying between the enemy's heavy cruisers and the beachhead. To do all this he had to sacrifice gunnery effectiveness, a small penalty because closer in his six-inch guns could not fire long enough to be effective before being put out of action themselves by the Long Lance. The destroyers, four under Burke at the head of the column and four astern under Bernard "Count" Austin, would be the untethered mad dogs charging in and inflicting the torpedo damage.

The three separated formations lost control so the results were mixed. Some of Burke's four ships scattered and he had to mill around reconcentrating them, at one point firing five-inch guns at Austin. Merrill kept his four light cruisers under tight rein and away from the danger of torpedoes. His ships smothered the enemy with shells and occupied his attention while the destroyers made their charge, doing what moderate damage they could. The American attack so bewildered the Japanese that they turned and got out, abandoning their mission. In the battle the Japanese ships were almost completely ineffective. The U.S. force had learned to survive by lightness afoot. Its gunfire wasn't much—Morison estimated that it achieved only twenty six-inch hits out of forty-six hundred rounds expended. Night spotting of gunfire proved to be terribly hard; the Japanese later said the Americans were consistently off in deflection. Merrill wisely declined to pursue, because dawn was coming and this time, for a change, it was the Americans who would be subjected to land-based air attack, from Rabaul.

At the Battle of Cape St. George, on 25 November 1943, Burke had the chance to use his patented tactic of hitting with the left and following with the right. He had five destroyers split into units of three and two, supporting one another five thousand yards apart. His nominally equal enemy had two new destroyers escorting three destroyer transports which were trailing thirteen thousand yards (a twenty-minute run) behind. The Japanese commander had no radar on, and his second force was, in effect, a protected, not a mutually supporting, force. Against Burke's force that was fatal. Burke detected the enemy on radar at eleven nautical miles and adjusted course, and fifteen minutes later at three nautical miles his three destroyers launched fifteen torpedoes undetected. They mortally wounded both lead destroyers. Then he took after the three destroyer transports. Bows to the enemy, his ships evaded a mess of torpedoes in a two-hour stern chase and sank one destroyer. It is fitting that the man who had conceived Moosbrugger's tactics could finish the Solomons night actions with his own little tactical masterpiece.

The Solomons: A Conclusion

How, then, to recapitulate a year and a half of night actions, eleven of which were distinguished as battles? Radar was the new sensor, it had to be integrated tactically, and nighttime provided a magnificent opportunity for so doing; black night was radar's element, and it should have given the United States a decisive advantage.

In the conditions under which these battles were fought, crossing the T meant very little. The best tactic was to approach on a broad front, bows on (short columns abreast in practice), wheel anywhere within range and fire a barrage of two or three dozen torpedoes, then point sterns toward the enemy's reply. One of the American errors was to forget that combat is two-sided competition. Line tactics were based on the strength of the broadside, which nominally had twice the firepower that could be unleashed end on. Line tactics overlooked the fact that a beams-to column exposed ten times as much hull to torpedoes as a line abreast pointed toward or away from the enemy. In force-on-force computations the U.S. Navy imposed on itself a fivefold penalty with the line ahead.

There was a no man's land of at least five miles in which no cruiser belonged. For it had come to pass that with a torpedo barrage, a handful of small ships could destroy more than their weight of the enemy and a superior force according to conventional reckoning. Somewhere in the Valhalla of warriors Jellicoe must have looked down on those dark nights punctured with the violence of the torpedo and with a thin smile shaken his head at the Americans who took so long to learn what he knew in 1916.

In the early battles the United States was foredoomed by pickup forces, thrown together and untrained, using the only tactics surface officers had practiced—those of the fighting column. Perhaps they were the only tactics the Americans were capable of executing at first, in the face of not only the Japanese threat but also the hazards of steaming at high speed in the darkness in mostly uncharted shoal waters.

Still, the impression remains that the early tactical commanders did not know better—did not know the importance of their radars and the dangers of a long column. Particularly in the early battles, they seem not to have had the sense of pace to keep control while the opposing forces were closing at speeds of a mile a minute. Tactical commanders sometimes drove their van destroyers until they were in danger of colliding with the enemy before opening gunfire. The Japanese never had that problem. Upon seeing an enemy, they pivoted and launched their spread of torpedoes. Their tactics had coherence. Before the war they built torpedoes into their cruisers, while the Americans took them out in the belief that all modern battles would be settled by guns outside of torpedo range.

From the outset the Japanese tactical commander was up front, usually in the lead ship, after the fashion of the great Admiral Togo. The American commander, in his cruiser flagship, was far back in his single column (in two battles as far back as the sixth ship). When fast action was required, maneuvering from the middle of the column was ineffective. The mess at Cape Esperance indicates the kind of problem that resulted. The placement of the American flagship in the middle was a consequence of tactical good sense that had become obsolete—an example of tradition reigning over an appreciation of new tactical circumstances. In the last five battles of 1943, including Empress Augusta Bay after Merrill turned his destroyers loose, the American tactical commanders were in front and the results were salutary. The Japanese practiced their night tactics in peacetime and knew how to fight from the beginning. To the Americans it became evident only through hindsight.

When I reread the details of these battles I expected to find that each American commander improved by learning on the job and training his units accordingly. Certainly the intense operations in the Slot—and a reader focusing just on the major battles misses the fact that ships were out night after night, patrolling, stalking, engaging in shore bombardment—helped to school and steel leaders and their crews in 1943. But with one exception we cannot find an American tactical commander who fought two night battles and improved. This is a commentary on the rapid turnover of ships and staffs. Already noted were the cases of Wright and Moosbrugger, who took command within forty-eight hours of battle. Admiral William F. Halsey, as commander in the South Pacific, never let pass an opportunity to fight and so was always scratching for ships, especially destroyers. No officer ever led more than two battles or commanded exactly the same set of ships. The single officer who fought two battles and improved was Arleigh Burke, and even he was not in overall command at Empress Augusta Bay. We may wonder how it might have been if Burke or Tip Merrill had fought as many battles as Nelson. Burke would have been the best prospect against a tactician he never had the chance to fight, the redoubtable Japanese wizard, the tenacious Tanaka.*

Old ideas of massing force came into question in the Solomons. A smaller force had the firepower at the prevalent short ranges to smash a larger force and survive. The commonplace principle of victory by superior concentration of offensive force had to be held in abeyance by the Americans. No doubt we will again see circumstances in which the wisdom of the Solomons prevails: when small ships armed with many missiles have the firepower to take out more than their weight of opposing force.

It is appropriate to conclude this section with an observation on strategy, specifically the strategic significance of the tactics in the Solomons. There appears to have been a tendency on the part of the Japanese to abandon winning positions (at Pearl Harbor, Coral Sea, Savo Island, Samar) and stubbornly to pursue losing causes (the whole

*In fairness it must be pointed out that the Japanese also fought with cruisers and destroyers that had not operated together. But doctrinal integrity and much practice in night operations kept them cohesive.

of the Guadalcanal campaign, Tanaka's dauntless reinforcement of Guadalcanal, the Battle of the Philippine Sea). The successful Japanese night tactics in 1942 were hit and run.* The successful American tactics in 1943 were hit and duck and hit again. American forces were sometimes punished because it was ingrained in them to stay when they came. But what was costly tactically may in the end have paid off strategically.

Radar and Air Defense

There is a coda to the Battle of Empress Augusta Bay that says much about the U.S. surface fleet's ability in 1943 to defend against air attack with radar. After the battle Merrill was well within striking range of the great Rabaul base and knew he would see a full-scale Japanese air attack in the morning, mirroring the American air attacks on the withdrawing Tokyo Express launched from Henderson Field in the Guadalcanal days.

Merrill gathered up his four light cruisers and four of his destroyers in their tight AAW wagonwheel. The attack came, one hundred strong, and if there was ever evidence that a modern surface fleet, concentrated and well handled, could deal with aircraft, this battle was it. Merrill's ships bristled with guns, for AAW defense was their business. The cruisers alone fired one thousand five-inch and over thirteen thousand forty- and twenty-millimeter shells. The Japanese achieved two bomb hits of minor impact and lost seventeen aircraft. The attack took seven minutes, which meant the cruisers fired about thirty-five shells a second, a withering barrage.† Radar had given Merrill early warning, radar fighter direction steered his inadequate land-based CAP, and (I am guessing) radar proximity fuses made his five-inch dual-purpose guns the most effective killers in the battle.

That was on 2 November 1943. Three days later Rear Admiral Frederick C. "Ted" Sherman's carriers *Saratoga* and *Princeton* were sent by Halsey to deliver air attacks on Rabaul. Up to this point

*This is not to deny Japanese courage in stopping to rescue survivors, which was done at great peril. Japanese tactics conformed to the prewar strategy of whittling the American fleet down to size.

†Morison, vol. 4, p. 321. Note that the cruisers fired over eight hundred shells for every enemy aircraft splashed.

Rabaul had been forbidden fruit. With at least seventy fighters on the ground, the harbor ringed with antiaircraft guns, and the firepower of seven or more heavy cruisers and a flock of lighter warships, the defense should have been overwhelming. But the Japanese were gearing up for a massive sweep, and Halsey, with not one heavy cruiser to fight a night action, felt he had to risk an air attack. He sent Sherman in to strike with forty-five attack aircraft and fifty-two fighters.* Japanese reconnaissance aircraft blew their report of the force, and Rabaul, without radar warning, was completely surprised. The American strike attacked in the tightest of massed formations and got out with losses, mostly in retirement, of only ten aircraft. They damaged four heavy cruisers, two light cruisers, and two destroyers, and the Japanese abandoned all thought of another night sortie to relieve Bougainville.

To see the significance of radar, one merely needs to reverse the locations and missions of the two fleets. American aviators had asserted their ascendancy over Japanese aviators, but we should not forget the roles played by U.S. warships in accelerating the process and the radar that was so effective in early warning and AAW gunnery.

Submarine and Sensors

This book says little about the submarine wars. That is because amphibious operations and air strikes required surface ships, and because fleet actions offered the best chance of controlling the seas. Submarines could deny but not exploit sea control. Submarines were spoilers, and still are that—except in nuclear war, in which case they are intended to play a central role.

Submarine roles are divided into support of the fleet and attacks on shipping in a kind of guerrilla warfare at sea. To support the fleet, submarines scout where other warcraft cannot go and attack and weaken the enemy. These were major roles in the German, Italian, British, Japanese, and American navies.

*Observe the fighter-heavy ratio. At Halsey's express orders, Sherman sent everything he had. His carriers were supposed to have been covered by land-based fighters.

The Philippine Sea (June 1944) serves as well as any battle to demonstrate the effectiveness of submarines in fleet support. Submarines first sighted the Japanese force and reported its general composition. Before the end of the battle they sank two big Japanese carriers, a better score than Mitscher's aircraft achieved. If one counts large, light, and escort carriers together, then during World War II aircraft of all nations sank twenty carriers of 342,000 aggregate tons. Submarines sank fifteen carriers of 306,000 aggregate tons. (At Midway the crippled *Yorktown* was actually sunk by the Japanese submarine *I-168,* but I credit the carrier's destruction to aircraft.) Surface warships sank two carriers, of 30,000 tons total.

In the submarine's guerrilla warfare against shipping there were three major efforts: the German campaign in the Atlantic, the British campaign to interdict German resupply of North Africa, and the American submarine campaign to isolate Japan from oil and other resources. By any standards all were formidable, and the last-named can be called successful. Even the U-boat campaign in the Atlantic, which resulted in unparalleled destruction of some of the bravest men ever to put to sea, can, as Admiral Gorshkov of the Soviet navy has pointed out, be called a strategic success because of the vast and disproportionate response imposed on the Allies, who had to expend many times more manpower and material to defeat the campaign than Hitler's navy committed to it. But then, barring certain inefficiencies on their part, the Allies had no choice. Nazi Germany had tremendous leverage and exploited it ruthlessly; the Allies needed control of the ocean's surface.

The submarines were the latest in a long tradition of raiders at sea. In the most rewarding of all guerrilla eras, Francis Drake and John Hawkins and their Elizabethan compatriots got double profit: they denied to the enemy the fruits of the ships they captured and took their booty to England. Later Raphael Semmes, captain of the Confederate ship *Alabama,* could destroy but rarely keep his prizes. By World War II surface raiders like the *Graf Spee* and *Bismarck* were doomed by cryptanalysis, aerial surveillance, and radar. To escape, twentieth-century raiders had to disappear under water. Throughout World War I and at the beginning of World War II submarines were surface raiders that submerged to evade attack. The significance of this was that U-boats forced under water by aircraft (far enough away from the convoys) were ineffective. In the middle of 1943 it was aircraft that broke the back of the peak effort by U-boats, when on an average day 104 of them lurked the seas. The Allies won the Battle of the Atlantic by a combination of offensive air patrols in the Bay of Biscay and defensive air patrols around convoys. Aircraft with radar first slowed U-boats in transit and then drastically curtailed their maneuvers in the vicinity of the convoys.

Radar was essential to the Allied effort. And the search radars stimulated one of the first big measure-countermeasure duels in the scouting business. We cannot recount here how the British kept ahead of German detection devices by changing frequency, but it is an instructive story. British operational researchers learned how to gather data that told them whether the U-boats were on to Allied radar frequencies.* In fact such analysis was not necessary. The innermost circles of Allied command knew already because the British had cracked the German code.†

Exploitation of the U-boat cipher—the Ultra secret—was the most important weapon of sensory warfare in the Battle of the Atlantic. Admiral Dönitz directed his U-boats from ashore. By 1942 wolfpacks were being used to scout and concentrate attacks on convoys. Since the U-boats could not communicate freely, Berlin played the role of tactical coordinator, opening up an opportunity for the greatest of all tactical signals exploitation. In May 1941 the British had pilfered a German cipher machine from the *U-110* and started to read (intermittently) the enemy's signals. In addition, the Allies triangulated U-boat positions with RDF to supplement cryptanalysis. German submariners thought that very short, technologically sophisticated

*Blackett, pp. 222–23. Among the several accounts of this electronic duel, one of the best and most concise is in Tidman, pp. 75–80.

†By 1943 American operations analysts had deduced this. As Tidman reports, Jay Steinhardt calculated that locations purportedly based on RDF (radio direction finding) fixes were ten times more accurate than analysis showed they should have been. He took this puzzle to Philip Morse, head of the ASW operations analysis group, who confronted his naval boss with the data. The cat was out of the bag, and Morse and Steinhardt, at least, were told the truth. Yet there was never a whisper of the secret in the vast outpouring of operations analysis literature.

burst transmissions would be impossible to use for triangulation; they were wrong.

The deciphered information was of the highest strategic importance, of course. Intelligence gave an exact count of the U-boat fleet, in both the Atlantic order of battle and the Baltic Sea during shakedown training. The code breakers provided bountiful information about the movements of U-boats and their plans of attack, ordered from thousands of miles away in Berlin.

> Another incalculable benefit of having been able to read Hydra [the cipher initially used by all operational U-boats] for so long was the insight which it had given us into the way the U-Boat war was being conducted, and perhaps even into the way that Dönitz's mind worked. We knew the U-Boat's methods, the average speed of advance when proceeding to and from patrol, the endurance of the various types of U-Boat and characteristics of their many commanding officers, the types of patrol lines favoured and the exact meaning of the short signals used for making sightings, weather or position reports.*

No wonder that superb American tactical study, *ASW in World War II,* published in 1946 and at the time classified, is so rich in detail—it even includes the names of German Aces.† The wonder is that the secret of Ultra was kept so well. In this official study virtually the only mention of code breaking is in connection with the capture of the *U-505* in June 1944. The authors write that it gave the Allies important information about the German codes. The truth is that the USS *Guadalcanal* and her escorts were able in quick succession to dispose of four U-boats, one of which was the *U-505,* because the Allies already possessed the code and knew where to send Admiral Daniel Gallery's hunter-killer group.

From what we know now about the influence of code breaking in the Atlantic and the Pacific it is reasonable to infer that:

— A guerrilla campaign at sea—a modern *guerre de course*—that is not covered will fail because of the capabilities of modern surveillance.

*Beesly, p. 116.
†Sternhell and Thorndike, pp. 4, 10, 11, 20, and 81.

— Due to overconfidence in high places, signals from shore to sea are especially vulnerable. Operational command from afar, which is tactical and reveals battle information, requires special signals discipline.

Tactical Interaction Between Land and Sea Forces

We come to the last important factor regarding sensors, scouting, and countermeasures. As emphasized in chapter 1, the *strategic* interplay of events ashore and at sea has always been the major determinant of the scene and scale of most naval battles as well as the opponents' aims. A major development in World War II was the added importance of *tactical* interaction, owing primarily to the new role of aircraft. In this brief section let us consider the effects the sensory revolution, which opened so many new possibilities, had on the information war and on naval command.

For the first time we see how tactical command was exercised from ashore to a significant degree: Yamamoto, Dönitz, Nimitz, and Halsey all participated to greater and lesser extents in the battle movements of their forces. In the record of Halsey's signals from ashore in Noumea may be found orders to commence South Pacific operations at explicit latitudes and longitudes at specific times. Tactical command and, even more commonly, what in current parlance is called operational art were exercised from ashore so that the striking force of ships and aircraft at sea could keep radio silence until their presence was discovered by the enemy. Even when cryptanalysis and RDF could not be exploited, traffic analysis of the volume of signals revealed impending operations. Often when the text could not be decrypted the address headings could, revealing the commands and ships involved.

The Japanese and Americans both used land-based reconnaissance aircraft, partly to help conceal the location of ships and aircraft at sea but also because land-based patrol aircraft had very long range and endurance. The Japanese, more than the Americans, used external reconnaissance to husband carrier assets for the attack. In the Mediterranean one of the most serious deficiencies, which sapped the Italian navy's confidence and morale, was the failure of land-based air reconnaissance.

I hardly need to comment again on the new reach of sea-based air attacks against shore targets. What might be noted is the fate of land-based attacks against targets at sea. Both the Italian and American air forces were supposed to attack warships at sea, and they largely failed. The Italian fleet, having been denied a naval air arm by Mussolini, had to depend on reconnaissance by the Italian air force and was grievously crippled by inept support. The Japanese, however, flew *naval* aircraft from fields ashore with success, most notably sinking HMS *Repulse* and HMS *Prince of Wales* with torpedoes. There is nothing inherently wrong with attacks launched from ashore against warships if the aircraft are trained and armed for the mission. Lack of mobility and the power to concentrate were the tactical constraints on the strike effectiveness of land-based aircraft in World War II. Weak command structure and neglect of the special training required to hit maneuvering warships were the main (and unjustifiable) reasons for the widespread failure of air strikes from land to sea.

Land-based maritime patrol aircraft proved highly effective against submarines because they could sortie on solo patrols safely and at long range. Due to the nature of their tactics they could be large and plodding. The British were slow to see their possibilities. After HMS *Courageous* was sunk by the *U-29* in the first week of the war, the Royal Navy failed to exploit the fact that land-based aircraft were a safe way to fly against U-boats. For three and a half years the British demurred, until the exigencies of the war in the Atlantic drove them to it. The turning point came in early 1943 with the transfer of bomber command squadrons to coastal command. At the same time the British persuaded President Roosevelt to earmark new U.S. Liberators, then starting to come off the production lines in large numbers, for ASW patrol, a mission that suited the characteristics of those remarkable, long-ranged aircraft.*

Nazi Germany missed a golden opportunity to exploit land-based aircraft at sea. After the fall of France, a few sorties by the Luftwaffe demonstrated that its medium-range bombers could attack Allied Atlantic convoys effectively. But Hermann Göring's penchant for attacks on land targets eliminated the possibility of developing and

*Blackett, p. 227.

committing German aircraft to shipping attacks in great numbers. That Germany might wake up to the opportunity was a worry haunting the harried Royal Navy through much of the war.

A separate book is needed to put amphibious operations in perspective. They have usually and rightfully been studied for their strategic content. But unquestionably the reach and range of weapons and the maneuverability of ships and aircraft changed the nature of amphibious assault between the Napoleonic Wars and World War II in fundamental ways that are almost impossible to exaggerate. Amphibious landings since World War II have demonstrated the growing land-sea interface and have made use of more new tactics. We may recall the stunning operation at Inchon, the landings at Wonson that were confounded by minefields, the recent event in which the British destroyer *Glamorgan* was struck by land-based missiles from East Falklands Island, and most recently America's swift use of aircraft and warship mobility in the taking of Grenada.

The multifaceted growth in the potential of land- and sea-based forces to operate against each other deserves and has received careful study. Not to be underestimated also are the roles of scouting, communications, the control of forces, and the countermeasures against them that have risen in importance and will continue to do so. The sensory revolution was the cause of all this. The growth of land-sea tactical interactions and of sensor technology are two of the great trends in tactics.

6

The Great Trends

On the Principles of War

In one of his best-known quotations Alfred Thayer Mahan wrote that "from time to time the structure of tactics has to be wholly torn down, but the foundations of strategy so far remain, as though laid upon a rock." Mahan thought the principles of strategy were easier to discern in history than the principles of tactics, because the latter "using as its instruments the weapons made by man shares in the change and progress of the race. . . ."*

Whether or not the foregoing is true, there is a distinction between military principles—Mahan's or anyone else's—and the actions that derive from them. Tactics change, but that does not preclude the search for tactical principles, and if there are strategic principles, that does not mean that strategies do not change. Strategies as well as tactics are influenced by "weapons made by man." We may forgive Mahan for not foreseeing how weapons of the future would influence strategy, but there was evidence of change even as he wrote. All the strategic practices of blockading were modified by the transition from

*Mahan, pp. 8, 88–89.

sail to steam. Sailing ships that stayed on station for months were being replaced by ships that lacked endurance and depended on coaling stations, the competition for which itself had a profound influence on strategy.

Throughout the hierarchy of enlightenment, from Truth, which the epistemologists say exists but is never known with certainty, to principles, which express our contemporary vision of Truth, to policy and doctrine, which are programs for concerted action based on principles, and finally to strategic or tactical decisions, which are individual actions guided by policy and doctrine—throughout this hierarchy, error creeps in. Mahan's skewed prognostications of World War I strategy, based on his own principles, illustrate as well as anything the margin for error.

The armies of the world are much keener to study principles than the navies have ever been, a condition Mahan deplored and one that has not changed much. Most students of military history have avoided the problem of differentiating strategic and tactical principles (and underlying logistics) by citing principles of *war*. Their search for enduring truth has ended in the compilation of many transitory lists. In an unpublished paper Captain S. D. Landersman assembled twenty-three lists of principles of warfare, including some essayed by naval officers.*

Principles on any subject have exceptions. The exceptions to the principles of war tend to be crucial. "In war every problem, and every principle, is a duality," wrote Liddell Hart. "Like a coin, it has two faces. This is the inevitable consequence of the fact that war is a two-party affair, so imposing the need that while hitting, one must guard."† Clausewitz, who made no lists of principles himself but who was the father of lists, confounded his readers with propositions and variations, some of which were counterpropositions. He is the practitioner's delight and the theorist's frustration. On the subject of strategic concentration of forces he wrote a very short chapter that can be quoted in its entirety:

> The best strategy is always to be very strong; first in general, and then at the decisive point. Apart from the effort needed to create military strength, which does not always emanate from the general,

*His lists are included in appendix B.
†Liddell Hart, p. 329.

> there is no higher and simpler law of strategy than that of keeping one's forces concentrated. No force should ever be detached from the main body unless the need is definite and urgent. We hold fast to this principle, and regard it as a reliable guide. In the course of our analysis, we shall learn in what circumstances dividing one's forces may be justified. We shall also learn that the principle of concentration will not have the same results in every war, but that those will change in accordance with means and ends.
>
> Incredible though it sounds, it is a fact that armies have been divided and separated countless times, without the commander having any clear reason for it, simply because he vaguely felt that this was the way things ought to be done.
>
> This folly can be avoided completely, and a great many unsound reasons for dividing one's forces never be proposed, as soon as concentration of forces is recognized as the norm, and every separation and split as an exception that has to be justified.*

The principles of war must have important exceptions, because most of the lists are in collective conflict, and few lists present their principles in order of priority. This is as it should be. Principles in conflict (concentration and economy, security and surprise) create a dynamic tension that is one small safeguard against heedless conformity; and the absence of priority makes the list memorizer pick and choose. Lists of principles help reduce the entropy of war. The danger lies in the reader taking lists as a substitute for study. A senior officer can hardly do more mischief to young officers than to lecture on a list of principles. He should lecture on their content.

The second weakness of principles of war is that they are usually reduced to key words, such as *concentration*. A word is not a principle. A principle is a statement of general truth. "Concentrate force" is a minimal statement of minimum value. "Mass superior force on a portion of the enemy's" is more specific and therefore more valuable. "Concentrate combat power at the decisive place and time to destroy the enemy, but do not mass your force so that it is vulnerable to enemy firepower" is still more specific, but it does not hold historically, even though it may be suitable today. Then there is the compelling way the same principle is expressed in Soviet theory: "Concentrate the main effort and create superiority in forces and means

*Clausewitz, p. 204.

over the enemy at the decisive place and time." Is that what we mean as well? At the very least, to be useful a principle of war must be current and prescriptive, clearly implying what actions must be taken to put it into practice.

A third weakness of the principles of war is that, uninterpreted, they do not distinguish strategy from tactics (or, if one insists, from "operational art"). Useful interpretations can be made, but in the transition from the general to the specific there is room for error. The Soviet military laws of war and "law-governed patterns" are very cogent, but they are also very abstract. The system of military science in the Soviet Union is one of the tightest ever constructed: confidence in theory and faith in determinism are between them a key to interpreting Soviet military plans and forecasting Soviet military actions.

Another problem with the principles of war is that they fail to distinguish between land and sea combat. Regardless of the trend (still to be discussed) toward greater tactical interaction between land and sea forces and the fact that the *strategic* influence of sea and land forces has always existed and not changed much, on the battlefield there are salient differences between military and naval principles of action. To illustrate, the following list shows statements by T. N. Dupuy on army battle next to my own statements on navy battle. The left-hand column is taken from chapter 1, "Timeless Verities of Combat," of Dupuy's recent book, *Understanding War*. The right-hand column is what I believe to be the naval counterpart of Dupuy's statements.

Land battle	*Sea battle*
1. Offensive action is essential to positive combat results.	This is true of sea battle.
2. Defensive strength is greater than offensive strength.	Defense is usually weaker.
3. Defensive posture is necessary when successful offense is impossible.	Defensive posture is inherently risk-prone and subject to incommensurate losses.
4. Flank or rear attack is more likely to succeed than frontal attack.	Attack from an unexpected quarter is advantageous, but the concept of envelopment has no parallel with land tactics.

Land battle (cont.)	*Sea battle* (cont.)
5. Initiative permits application of preponderant combat power.	The power of initiative is especially valuable at sea.
6. A defender's chances of success are directly proportionate to fortification strength.	Defensive power is solely to gain tactical time for an effective attack or counterattack.
7. An attacker willing to pay the price can always penetrate the strongest defense.	This is true of sea battle, given the wherewithal.
8. Successful defense requires depth and reserves.	At sea, setting aside reserves is a mistake.
9. Superior combat power always wins, if one takes into account the value of surprise, relative combat effectiveness, and the advantages of defensive posture as elements of strength.	When the appropriate qualifications are considered, it is possible to say that superior force will always win at sea. However, it is better to say that when two competitive forces meet in naval combat, the one that attacks effectively first will win.
10. Surprise substantially enhances combat power.	This is true of sea battle.
11. Firepower kills, disrupts, suppresses, and causes dispersion.	This is true of sea battle.
12. Combat activities are slower, less productive, and less efficient than anticipated [from peacetime tests, plans, and exercises].	While this is often true, there are many examples of naval engagements in which the results come more swiftly than expected. Perhaps there is less friction at sea than on land.
13. Combat is too complex to be described in a single simple aphorism.	This is true of sea battle.

Though they do not consistently apply to sea combat, I have no reason to quarrel with Dupuy's military verities. Principles of war are useful. Like all good theory, they help explain the whys and wherefores (in contrast with practice, which deals with whens and wheres and hows). And yet ultimately there is dissatisfaction with principles because they lead nowhere. After diligently distilling the statements of the wisest authorities into a flawless list—if it could be done—one would be positioned at an intellectual cul de sac. Without much loss, one might as well say that the contribution of principles to warfare boils down to outthinking the enemy. Understanding the processes of combat is a better approach to tactics. Processes are the navigator's science and art, principles the stars he uses to find his way.

The Processes of Combat

The key to fruitful study of tactics is an appreciation of how battles transpire in time and space. The activities—the dynamics—of combat are the wellspring of understanding. *Dynamics* suggest time-dependent models—descriptions of combat processes. Generally, models are images of processes; specifically, they are mathematical models, simulations, and war games that probe and add detail until the battlefield itself becomes the ultimate laboratory of understanding. That the study of battlefield dynamics constitutes the proper approach to study is reinforced by the universal terminology of warfare; *power, potential, energy, pressure, mass, momentum, movement,* and *force* refer to the dynamics of physical bodies that warriors apply to the processes of combat.

Looking at fundamental processes, we will be able to reexamine history and frame some conclusions about trends, constants, technology, and battlefield contexts. We will see how technology wrought a change in the way each process was executed. For example, we have already observed how the first process, the delivery of firepower, changed when the age of sail gave way to the age of the big gun and the latter gave way to the age of air power. Technology changed the delivery process and with it the tactical manner of concentrating firepower. Evolving trends will afford clues to future combat in what is called the missile age.

The abstractions below do not follow from history like laws of physics from a falling apple. Evidence for social phenomena is seldom as conclusive as it is for physical phenomena. But it will have to serve.

Naval tactics are built on five propositions, each of which involves a process:

— Naval warfare centers on the process of attrition. Attrition comes from the successful delivery of firepower.
— Scouting—locating the enemy sufficiently to deliver effective firepower—is a crucial and integral process of tactics.

— C^2 is the process that transforms scouting and firepower potential into the reality of delivered force.
— Naval combat is a force-on-force process tending, in the threat or realization, toward the simultaneous attrition of both sides. To achieve victory one must attack effectively first. Therefore actions taken to interfere with the enemy's firepower, scouting, and C^2 processes are also of fundamental importance.
— Maneuver is also a tactical process. In fact, maneuver in battle was once the classic definition of tactics. Maneuver is the activity by which C^2 positions forces to scout and shoot. Battle maneuver deserves—and receives—attention, but in the structure that follows it is an orphan, no longer warranting its earlier status.

We can also view firepower delivery, scouting, and C^2 not as processes but as functioning elements of naval forces: firepower, scouts, and C^2 systems. Confronting these elements of force are three opposing elements: counterforce, antiscouts, and C^2CM systems.

Firepower and Counterforce

Firepower is the capacity to destroy the enemy's ability to apply force. Counterforce is the capacity to reduce the effect of delivered firepower. We could refer to offensive and defensive power, but it is a useful cue to retain the asymmetry of counterforce as the defender's response to firepower. Though the practice is less common today, navies historically answered enemy firepower by building survivability—called staying power in the days of sixteen-inch guns and eighteen-inch turret faces—into the hulls of warships.

Scouts and Antiscouts

Scouts gather information by any and all means: reconnaissance, surveillance, cryptanalysis, and every other type of what some call information warfare. But scouting is not completed until the information is delivered to the tactical commander. Scouts deliver tactical information about the enemy's position, movements, vulnerabilities, strengths, and in the best of worlds, intentions. Antiscouts destroy, disrupt, or slow enemy scouts. I would prefer to call this interference screening, but screening has come to mean both antiscouting and counterforce (that is, ASW and AAW screens put as much or more emphasis on countering an enemy attack as on reducing the quality of his information).

C^2 and C^2CM Systems

Command decides what is needed from forces and control transforms the need into action. These are processes. C^2 systems are defined, perhaps a bit artificially, as the equipment and organizations by which the processes are performed. Command is embodied in the commander, his staff, and their material resources, for example a simple maneuvering board or a complex display of processed scouting information. Control is embodied in communications equipment, the operation order, fleet doctrine, and the signal book. Command and control countermeasures (C^2CM) are steps to limit the enemy's ability to decide (command) and disseminate decisions (control). C^2CM devices include missiles that destroy command centers and flagships. More commonly they are communications jamming equipment. At its most subtle, C^2CM is carried out by espionage agents planting misinformation and (when used to confuse the tactician's decision rather than divert his weapons) false contacts. However, spies can perform acts of scouting, and signals exploitation (for example, by RDF) can be a form of scouting. C^3 countermeasures (C^3CM), a term frequently used, refers to actions taken against firepower, scouting, and C^2, sometimes all simultaneously. It is better to differentiate which enemy process is the principal target, and I shall be careful to do so when referring to counterforce, antiscouting, and C^2CM actions.

A tactical commander uses C^2 to allocate his forces for four activities: firepower delivery, counterforce delivery, scouting, and antiscouting. Meanwhile the enemy commander is doing the same thing. Many weapon systems, and all that are able to operate independently, such as submarines, will have some capacity to conduct all four activities. From a fleet commander's point of view, one of his major tactical responsibilities is to assign functions to his forces. He must also integrate the expected contributions of systems not under his command, such as national surveillance satellites, army surface-to-air weapons when he is in port, and air force interceptors that stand between his fleet and the enemy airfields.

We are now equipped to examine, first, the historical trends that have altered the character of each process and the tactics associated with them. This chapter concentrates on the causes and effects of tactical change. In the next chapter we will examine the historical constants—what tactics have not changed, or what tactics, such as surprise, have played invariant and reliable roles. Insofar as I can see, knowledge of trends and knowledge of constants are equally important.

However, separate analysis of each process—delivery of firepower, counterforce activity, scouting, and antiscouting—can be deluding. Concerted action wins battles. The processes must be coordinated by a tactical commander as the instruments of an orchestra are blended by a conductor. At the same time, an enemy commander is making decisions about his forces and the timing of his attack. Both sides take steps building toward combat's climax and outcome. Except in battles where the result is a foregone conclusion, the winner will be the fleet most united in seeking the opportunity to attack effectively first.

Maneuver

Maneuver has a unique place in the rollcall of processes. Through maneuver the elements of a force attain positions over time. When all elements are in good positions to execute their assigned functions, the best hope of victory is established. At sea, positions are not fixed geographically (though some fleet components may be at immobile sites ashore).* Positions of opposing fleets exist in mutual relationship, and as elements maneuver, their relative positions are in a complicated state of flux. In naval combat, the relationship between the range and bearing of one force and the range and bearing of the other is a paramount tactical consideration, which simultaneous maneuver continuously alters. The relative positions of one's own forces are also crucial and, even among skillful forces, sometimes seem as frustratingly difficult to track as the positions of the enemy.

Sailing ships stayed in a tight column for cohesion and sought a

*Ground combat seems different. Defendable positions achieved first have absolute value.

position to the windward or leeward of the enemy. Battleships maneuvered to cross the enemy's T. Picket submarines are prepositioned to scout and attack in waters through which the enemy may pass. Aircraft are put on a CAP station so they may be vectored or may maneuver of their own volition to apply firepower. Deck-launched interceptors are in a state of readiness that is predetermined to allow them time to move into a position for attack. In every case the emphasis is on the *timely positioning* of forces, which enables one to scout and shoot better than the enemy. Maneuver is the means to the intermediate end of establishing relative positions toward the ultimate (tactical) end of delivering firepower.* Particularly in modern naval combat, when weapon and sensor ranges so dominate ship and even aircraft maneuver, that is, the ability to shift position, maneuver should enter into the tactician's calculations as the vital feasibility check against his plans for positioning and timing.

Moreover, the physical speed of ships, aircraft, and weapons may easily be confused with the speed of decision and the speed of execution of decision. To eliminate ambiguity would be to eliminate an understanding of the interrelationships. The epigrammatic Sun Tzu is often quoted as saying, "Speed is the essence of war." It is apparent from his next sentence that he was speaking of command: "Take advantage of the enemy's unpreparedness." He was also speaking of mobility, for he then says that to capitalize on this unpreparedness one must move swiftly to the enemy's area of vulnerability. Mahan's aphorism, "The true speed of war is . . . the unremitting energy that wastes no time"—though he referred specifically to battle-line speed—is deliciously ambiguous. Mahan grasped as well as anyone that timely delivery of concentrated firepower involved the conjoining of everything: decisions and their dissemination, strategic aggregation, tactical positioning, and fast, accurate gunfire.

Maneuver and mobility are terms sometimes used interchangeably. Mobility we will take to be an element of strategy, operational art, or grand tactics. When Mahan described mobility as "the prime characteristic of naval strength," he was expressing a truth of strategy.

*Again I would not take the same stance regarding ground battle or tactics.

Mobility consists of the capacity:

— To move long distances in a relatively self-sustaining manner. Mobile logistic support forces—the "fleet train"—have made this possible.

— To move apace, that is, quickly relative to the movement of ground forces or the establishment of new airfields or missile bases ashore.

— To operate at length, up to months on or near station. Both naval bases and mobile logistic forces have permitted this.

For mobility, ships without mobile logistic support forces must have their own built-in range and endurance. Nuclear propulsion in surface ships is an attribute of strategy. In contrast, nuclear propulsion in submarines is an attribute of both mobility and effective tactical maneuver and stealth. The significance of mobility deserves elaboration, but it is not our subject.

Maneuver is *tactical* speed and agility. Fleet maneuver must be collective, coordinated motion, so it is impossible to divorce C^2 and speed of decision from this discussion. Tradeoffs can be made between the amounts of time it takes to scout, to assimilate information, to decide, to order, to maneuver, and to deliver effective firepower. Speed and agility of platform—warship or warplane—are two elements of speedy action. Warriors like a maneuverability advantage because it helps compensate for other shortcomings.

Two trends have appeared in maneuver today. The first is a shift of emphasis from speed of platform to speed of weapon. Until World War II maneuvering the fleet was the very heart of tactics. During the war aircraft speed took precedence over ship speed. Since the war missile speed and range have created a tactical environment in which weapons will be delivered without much change in ship position. Missile speed and agility are nullifying even aircraft agility, and farsighted combat aviators concede that missile maneuvers now dominate air warfare, as aircraft dominate ships' tactical manueuvers.

The second trend of maneuver is a corollary of the first. Ship maneuverability has diminished in importance and given way to scouting. "The fundamental tactical position," wrote Guiseppe Fioravanzo, "is no longer defined by the *geometric* relationship of the opposing formations, but by an *operational* element: the early detection of the enemy."* The significance of this development is that firepower might be more easily concentrated at long range, when naval forces are physically divided. This is a possibility emphasized by Admiral Elmo Zumwalt, Admiral Worth Bagley, and I believe, Admiral Stansfield Turner.† Since C² is also required for concentration without massing, I must defer further discussion until the appropriate section of this chapter.

I have alluded to the larger no man's land brought about by longer weapon range. It is harder now to distinguish between movement to the scene of battle (strategic mobility) and movement to attack effectively first and win (tactical maneuver). Since battles can occur at very long range, the value of mobility is perhaps diminished, certainly modified, and in theory very long, powerful missiles neutralize the strategic capacity of mobile forces to shift the scene of action. On the other hand, long-range delivery systems in effect endow speed with a new *tactical* attribute by allowing a ship to move out from under an enemy's targeting solution for his long-range missiles.

Historically maneuver has been used for three purposes:

— Advantageously concentrating offensive or defensive force. Except strategically, fleet motion has diminished in importance.

— Striking more quickly. Important as this element of tactics is and always will be, speed in ships and aircraft themselves is diminishing in relative importance.

— Protection by evasion of weapons. The value of speed in self-defense is increasing. Agility is still important, but hardly more so than in the past.

*Fioravanzo, p.209.

†These officers were prominent leaders of the U.S. Navy in the early 1970s. Zumwalt was CNO from 1970 to 1974, and Bagley was his vice chief during much of the same period. Turner's influence on the navy was probably at its peak when he was president of the Naval War College, from 1972 to 1974. They were creative, imaginative and—in the collective mind of the conservative navy—unconventional. They have continued to comment on current naval topics since retirement. None has recorded his tactical thinking publicly in a detailed way, but judging from the actions and shipbuilding aspirations of all three in office, I infer that they put tactical emphasis on concentration of force between numerous, widely distributed ships and aircraft linked by sophisticated communications.

Firepower

The best known trend in the history of warfare is the increase in weapon range, from two miles or so in the days of fighting sail, to fifteen miles or more in the era of the big gun, to three hundred miles during World War II, and to six hundred miles today. Since intercontinental missiles with nuclear warheads reach halfway around the globe, we have arrived at a plateau: in nuclear warfare the potential tactical battlefields are the United States and Russia.

Regardless of this trend, maximum weapon range has never been very interesting to the tactician. *Effective* range is what matters. The long guns of sailing ships were effective at only about three hundred yards, and carronades at an even shorter distance. Around 1900, before continuous-aim fire, it was estimated that a battleship would take fifty minutes to reduce an enemy to impotence at a range of twenty-five hundred yards. By 1914 (with good visibility) an enemy could be put out of action at ten thousand yards in ten minutes. The effectiveness of the major-caliber battleship guns at the time of World War I is depicted in figure 6-1 (which assumes good visibility, acceptable sea states, and the use of visual range-finding equipment). Fire control in that war was the key. The performances in figure 6-1 are nominal and take little account of the smoke of gunfire or destroyer smoke screens. At the Battle of Coronel, the Battle of the Falklands (1914), or the battle cruiser action at Jutland, when visibility was pretty good, the outcome would be decided outside ten thousand yards. Nevertheless, when visibility was an issue, for example in the action between the battle lines at Jutland, a fleet could save itself by opening out expeditiously, as the German High Seas Fleet did.

In World War II radar ranging changed that. Gun ballistics being highly accurate, with refined fire-control systems, even medium-caliber five-, six-, and eight-inch guns could be fired accurately almost to their maximum range.* The increase in effective range of naval weapons in the half century from 1898 to 1948 was about tenfold, an order of magnitude.

*We should never forget, however, that situations can arise such as those described in chapter 5 regarding gunnery effectiveness in the Solomons.

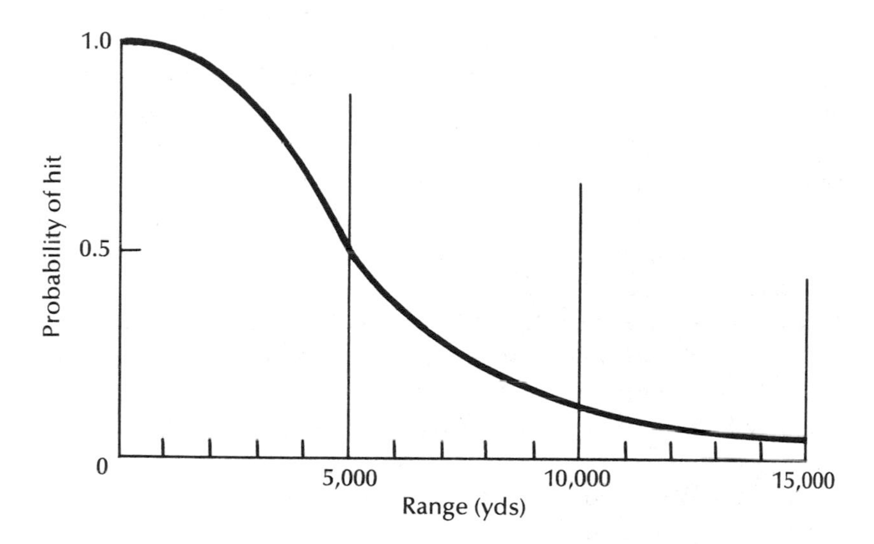

Fig. 6-1. Probability of Hitting with Battleship Guns as a Function of Range, ca. 1916
(*Source*: McHugh, table 4-2, p. 4-14.)

The increasing effectiveness of gunfire was obscured by aircraft early in World War II. But in aircraft themselves there was a distinction between gross weapon-delivery range and effective range. In the 1930s land-based B-17 bombers were designed with attacks on warships at great range especially in mind. Yet horizontal bombers turned out to be almost totally ineffective; they had difficulty finding naval targets at long range and even more difficulty hitting them at any range. Naval aircraft of much shorter ranges proved to be the best ship killers.

In a short unpublished paper A. R. Washburn of the Naval Postgraduate School makes a comparison between naval aircraft and naval guns. For representative battleships and aircraft carriers, he plots firepower versus range as depicted in figure 6-2, which shows the weight-rate of firepower delivered in eight-inch-gun equivalent rounds. For the USS *Iowa* and the Japanese ship *Yamato* the rate of the main battery is taken to be two rounds and one round per minute respectively, and for the aircraft, turnaround time is taken to be an hour. The carrier *Enterprise*'s whole "main battery" was paltry in comparative rate of delivery. The range of the ship's aircraft was decisive against enemy carriers because the weight of attack, in 1942 and 1943 at least, was *sufficient* for decisiveness. Before the war the tacticians' mental model looked about like Washburn's curves. The pro-battleship community doubted that the weight of an air strike would be sufficient for decisiveness, especially when defensive considerations were factored in. But the pro-carrier community tended to think of the weight of attack in terms of the attacking wave of bombers, sixty times as great as that the figure shows. An air-wing attack was to be a decisive pulse of power. Washburn also aggregated the firepower of the U.S. Fleet in 1939, the result of which is shown in figure 6-3. We may hardly wonder why carrier aviation's potential for decisive action was controversial before World War II.

The modern trend continues toward greater nominal range by means of compact ballistic- and cruise-missile propulsion systems, and toward greater effective range by means of sophisticated fire

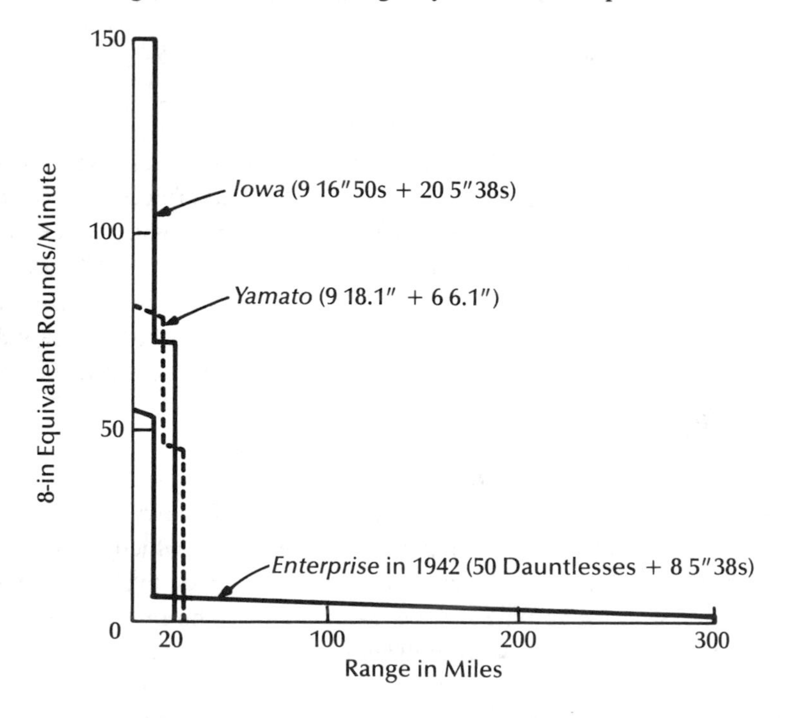

Fig. 6-2. Rate of Fire Versus Range in 8-in Rounds/Minute Equivalents. (*Source:* A. R. Washburn.)

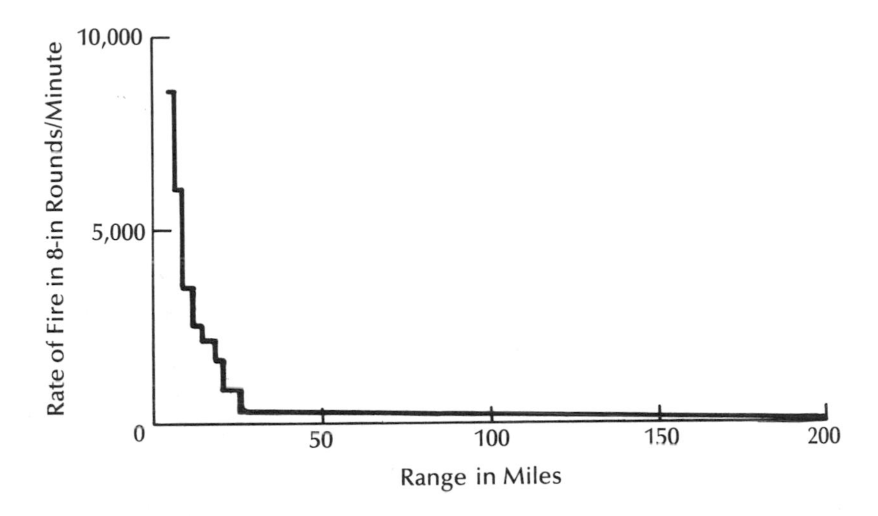

Fig.6-3. Rate of Fire of the Entire U. S. Fleet in 1939.
(*Source:* A. R. Washburn.)

control and homing systems. There is also a trend in the direction of raw destructive power. As important as range has been the growth in weapon lethality. This has been systematically studied for land-combat weapons by T. N. Dupuy, whose work on the increase of theoretical lethality is displayed in figure 6-4*. Notice that the vertical scale is logarithmic. After concluding that, even excepting nuclear weapons, the lethality of weapons has increased by five orders of magnitude—that is, one hundred thousand times—between the middle of the sixteenth century and the present time, Dupuy exhibits a paradox. While weapon lethality on the battlefield grew, the rate in personnel casualties per unit time shrank. Why? One prominent reason was the increased dispersion of troops on the battlefield.

Hanging over the head of civilization is the Damocles sword of nuclear weaponry. It is small comfort to observe that we have probably reached a final plateau in the increase in both the range and lethality of nuclear weapons. Speaking tactically, there is some uncertainty as to whether the *effective* range of these weapons is truly intercontinental, since they have never been employed. Civilization, which must suffer the consequences of badly aimed weapons, and tacticians, who must deal with the uncertainties of ineffective delivery,

*Dupuy (1979). See especially p. 7 and chapters 1 and 2.

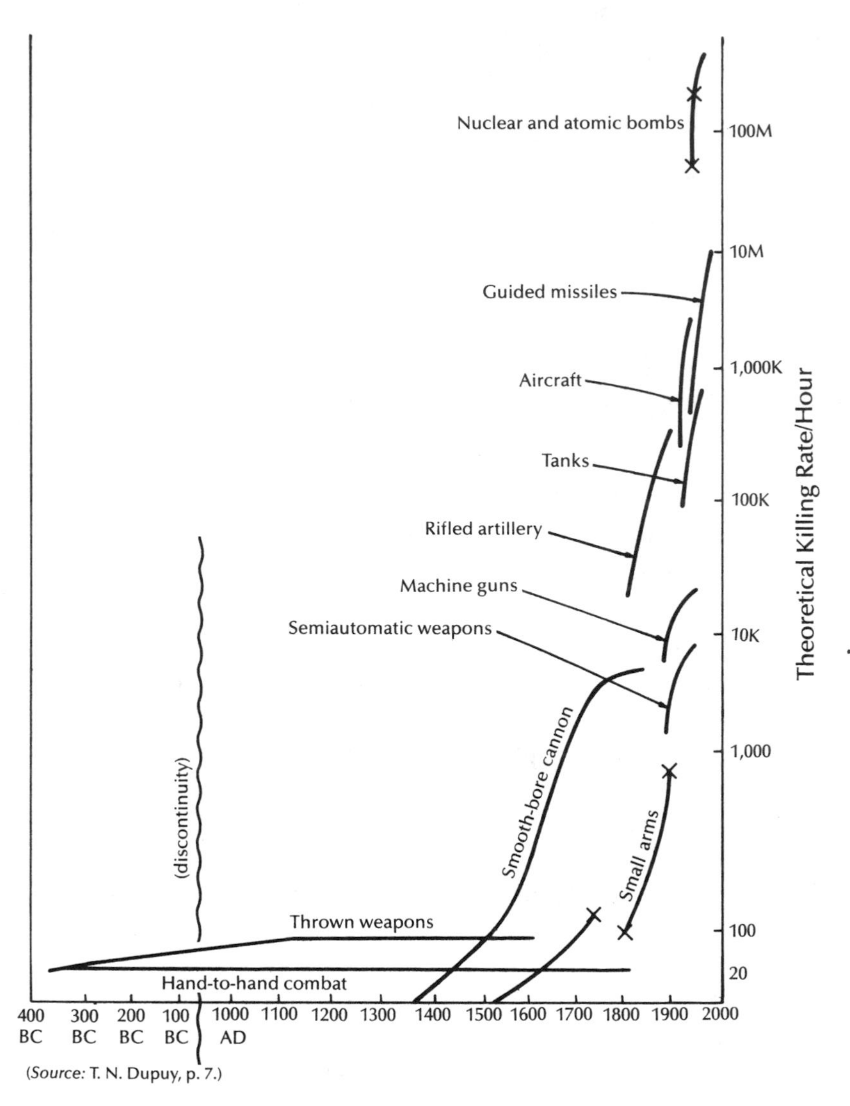

(*Source:* T. N. Dupuy, p. 7.)

Fig. 6-4. The Increase of Weapon Lethality over History

may both hope *for* missiles, ballistic or cruise, that will travel thousands of miles and land as predicted, within a few score meters. Geonavigational accuracy, measured in meters at both the sending and receiving ends, is necessary for tactical effectiveness at intercontinental ranges.

For the naval tactician, the use and the lurking threat of the use of nuclear weapons create enormous practical difficulties. Still, we may conclude that the trends of increased effective weapon range and lethality presage:

— A change in the form of defense. This will be taken up in the next section of this chapter.
— A further erosion of the distinction between land and sea battle, to be taken up in chapter 9. The emphasis here is on the greater potential for combat between land and sea forces; this is an important *tactical* trend that heightens the struggle between land forces, which have greater recuperative power, and sea forces, which are less easily targeted because of their maneuverability.
— For worldwide nuclear war, the fusion of strategy and tactics. This will be taken up and laid to rest at once.

Because of the unpleasantness of the prospect and a general trust in the nuclear umbrella's ability to deter, too little thought has been given to the combat execution of intercontinental war. The term *strategic weapons* itself is beguiling: if a general nuclear war breaks out, the world is in effect a tactical battlefield, directed by commanders and staffs operating in the highest-level command posts. It is therefore crucial that tacticians direct attention to this area.

In the 1960s and 1970s the paradigm of nuclear war was this: the president pushes the Red Button, which dispatches thousands of Minuteman and Polaris missiles in one huge spasm. One does not have to take that always flawed and now obsolete model literally to see the need for a more effective, not to say rational, method of C^2 at the end of the twentieth-century. In such a war of supreme destructiveness the centers of C^2 themselves would be attacked with unprecedented ferocity. Commanders and their staffs in positions up to and including unified and specified commanders would be the officers in tactical command, selecting targets, gathering intelligence, issuing orders to fire, and conducting the battle on a worldwide scale. On these commanders and staffs devolve all of the usual tactical responsibilities—for operational plans, for communications with forces bearing arms, for timeliness of commands to move and shoot, for access to survivable scouting information, and for tactical training in the heat of simulated battle. Even the commander in chief of the Pacific, a four-star admiral, may not think of himself any longer as dealing solely with strategy and logistics. When he controls the targeting and release of nuclear weapons, he will be a tactical commander on the field of battle.

Counterforce

The prominent trend in defense is away from survivability through armor, compartmentation, bulk, and damage control, and toward cover, deception, and dispersion. To discuss defense a distinction must be made between "dumb" weapons—shot, shells, and bombs—and "smart" weapons—manned aircraft and guided missiles. For purposes of analysis, aircraft are themselves rather like guided missiles, except that they may be used more than once.

When shells, torpedoes, and bombs dominated war, there was every possibility of building substantial staying power into warships. There were limited means of achieving cover and deception since battles were fought within visual range. The smoke screen was the most common cover device. Bombs from horizontal bombers could usually be evaded, and salvo chasing had some temporary effect.

Torpedo countermeasures were the prototypes of modern defenses against missiles. A torpedo was highly lethal and the best defense was to avoid it. To compensate, attackers fired spreads. Submarines tried to fire them at point-blank range, and surface ships tried to coordinate simultaneous attacks in the greatest numbers possible. In every case, brave attackers closed as much as they dared, because torpedoes, once fired, were irretrievable. There are many parallels between the torpedo and the modern missile scenarios. Missiles, for example, should also be defeated before they hit.

Even in their heyday armor and hull strength were hardly ever thought of as offering as much security against shells as deep bunkers in the ground. Armor was a dilatory device, used to forestall enemy firepower until one's own offensive power took effect. In those days there was much discussion of the division of a ship's displacement between firepower, staying power (protective armor), and propulsion power. Before and after World War I each country had its own style: Americans sacrificed speed for guns, armor, and radius of action, the Germans opted for staying power, the Italians emphasized speed, and the British (rather like Americans today) incorporated habitability for extended worldwide deployments in big ships.

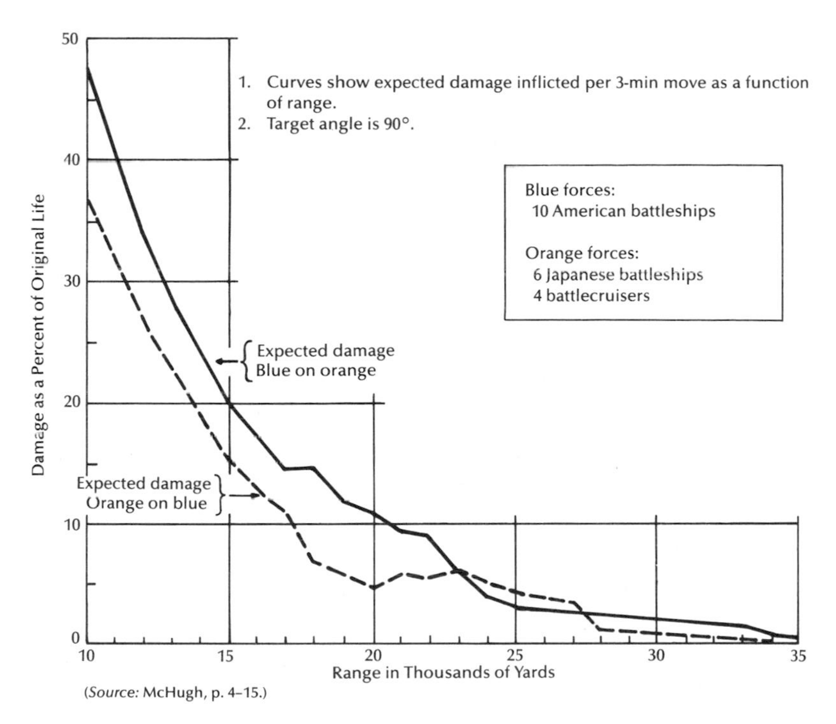

Fig. 6-5. A Naval War College Comparison of Opposing Battle Lines, 1926

In the war games of the battleship era the typical first-line dreadnought had a life of about twenty major-caliber hits, the predreadnought a life of twelve hits.* The loss of firepower and maneuverability was treated as a nonlinear function of the number of hits; that is, a dreadnought suffering ten hits in U.S. war games would lose more than half its firepower and speed.

In the 1920s game experts at the Naval War College saw that it was possible to aggregate the effectiveness of the battle line, taking both firepower and staying power into account. Figure 6-5 displays a comparison between opposing battle lines on parallel courses, with all ships' broadsides able to bear and with (mark this) unimpeded

*Around the time of World War I, one or two torpedo hits were estimated to be fatal.

visibility. In the figure the U.S. Pacific Fleet (blue) comprises ten battleships. They are more heavily armed and armored than the six battleships and four battle cruisers of the Japanese (orange). At fifteen thousand yards in *three* minutes the Japanese line will lose twenty percent of its original life, the American line only fifteen percent. The relative Japanese strength will deteriorate very rapidly. If the Japanese are to have hope of winning, they must count on their speed to cross the T before the range is closed, and they must try to maintain an interval between battle lines of about twenty-five thousand yards, where they have an advantage (according to the figure).

As suggested in chapter 4, both the Americans and the Japanese knew these relationships in the 1920s. American fears centered on the Japanese advantage in line speed (twenty-three knots against eighteen knots), the possibility of surprise, and the lurking danger that the U.S. Fleet would be too crippled after eliminating the Japanese to fulfill its mission (in war games, the relief of the Philippines). The Japanese hoped that their submarines would inflict initial damage, their aircraft and Long Lance torpedoes further damage, and that their *Mogami*-class light cruisers, retooled secretly with eight-inch guns, would significantly augment the battle line. There were, as we know now, catastrophic surprises to both sides after the war in the Pacific commenced. It is useful, nevertheless, to recall once more the coherence of planning on both sides, the legitimate American concern for the unexpected, and the lightning pace of decision. The kinds of force comparisons I have shown were familiar, and the games were played through to add dynamics to static plots such as that presented in figure 6-5. Though the pace of destruction rarely unfolded as rapidly as that figure foretold, still it was furious, and we may believe that the friction of war was not wholly discounted when these little planning aids were used.

In World War II defensive *weapons* assumed unprecedented prominence. By 1942 a flood of AAW weapons was being installed, with radar sensors, deadly proximity fuses, and new, capable fire-control systems to lead and hit fast-moving targets. By 1944 attacking aircraft faced a veritable curtain of fire. In the last year of the war, modern surface combatants had redressed the balance of power they had lost to naval aircraft.

The ascendancy of the ship lasted a mere moment, for at the end of World War II it was eclipsed by the atomic bomb. One bomb delivered on target would sink one ship, maybe more, and armor was obsolete against it. Cover and deception and the urgency of a first strike took on overwhelming significance. Haystack dispositions were spread over hundreds of miles, designed to deceive and forestall attacking bombers and submarines until the offensive strike was in the air. Air interceptors, AAW missiles, and ASW weapons were more than ever temporizing weapons. The American posture was all the trickier because the U.S. Navy could never attack first, certainly not with nuclear weapons. How to buy enough time to deliver a massive strike ashore was the tactical question. Judging from the enormous Soviet naval effort to counter U.S. carrier task forces, the Americans were eminently successful. But they paid a price: with nuclear war in mind they built ships without much survivability against conventional munitions. They concentrated on long-range defensive weapons—air interceptors and missiles—and neglected the guns and the modern close-in "point" defenses that were analogous to the twenty- and forty-millimeter guns of World War II. They also neglected the development of new soft-kill devices, short-range systems that could not reach out far enough against nuclear weapons. The Royal Navy followed a similar bent and neglected damage control and point defense. It suffered the consequences when its ships fought to retake the Falklands with conventional weapons.

In the preceding section on firepower I mentioned T. N. Dupuy's paradox: Although weapon lethality has increased one-hundred-thousand-fold since the sixteenth century, the casualty rate in ground combat has diminished. No corresponding data exists for sea combat, so let us see what can be gleaned from Dupuy.

First, the number of highly capable weapons per soldier on the battlefield has decreased; weapons like tanks, fighter-bombers, and heavy artillery account for much of the rise in the theoretical killing rate. Yet one set of Dupuy's data shows that in modern battle a greater precentage of casualties has sometimes been inflicted by other than the most capable weapons: infantry small arms exceeded artillery in producing casualties *after* the range and lethality of artillery rose dramatically. Often the second-best weapon performs better because the enemy, at great cost in offensive effectiveness, takes extraordinary measures to survive the best weapon.*

Second, in the past, ground combat weapons often could not be aimed at their targets when the latter took cover, and therefore they were less than nominally effective. As weapon range increased, the effects of weapon delivery inaccuracy increased greatly. Area fire breaks up concentration of force and suppresses enemy fire, but it is inefficient for killing.

Third, increasingly troops have been dispersed for survival. Dupuy estimates that between the Napoleonic Wars and the 1973 Arab-Israeli War the average density of troops on the battlefield was diluted by a factor of two hundred.†

Of special relevance to naval warfare, Dupuy observed that the rate of destruction of hardware, tanks in particular, far exceeded that of troops. A highly dangerous enemy machine draws fire because it is dangerous, but it is not correspondingly manpower dense and so not many casualties result from its being put out of action.

Ground forces disperse over the battlefield not evenly but in clusters. A platoon of men is a small cluster of force, a tank is a big one. Ships at sea are still bigger clusters of force, whether measured in firepower, manpower, or dollar value. The smallest unit that can be dispersed is a ship. When dispersion is an important means of defense, small ships and distributed firepower are an important advantage. Much of the modern debate over the size of warships concerns the comparative merits of dispersal in small ships (to complicate enemy targeting) and of concentration of force in large ships (to fight off the enemy). The Pacific carrier wars provided some information that

*We saw this phenomenon in the Falklands War. The Argentine air force lost only eight, or about ten percent, of its aircraft to the British ships' most expensive AAW defense, their SAM missile batteries (Sea Darts). The Argentine pilots knew that if they hugged the water the SAMs would be ineffective, and the British ships shot down most of the attackers with short-range weapons. Nevertheless, the "ineffective" SAMs were vital to the defense because they constricted the Argentine pilots' maneuvering room, helped make the British close-in defenses more effective, and forced the pilots to drop their bombs at so short a range that sometimes the ones that hit had had no time to arm.

†Dupuy (1979), figure 2-4, p. 28.

illuminated the problem. It was *defensive* fighting power that decided whether a force should mass or disperse. Today if a commander's fleet comprises many ships with strong defenses he masses and fights the enemy off. If he has few ships or weak defenses he must disperse. In either case he is buying time to carry out his mission, which is not to steam around waiting to be sunk. If the defense cannot buy time for the offense to perform, then the fleet ought to be somewhere else.

American experience since World War II has been in conventional war, with U.S. warships for all practical purposes operating in a sanctuary, either out of range of enemy weapons or fighting an enemy (notably North Korea and North Vietnam) that lacked the force to attack effectively and feared the consequences of a counterattack. The U.S. Navy's bread-and-butter missions for thirty years, these "projection" operations have probably bred complacency about the nature of combat, which is not always so one-sided an affair; they have certainly affected American attitudes toward defense, damage control, ship construction, and survivability.

The ability to sustain damage and continue fighting in conventional war can be built into modern ships. In the unclassified volume of the report on the Falklands War by the Department of the Navy, there is a reminder of this point.

> The EXOCET missile that sank *Sheffield* [the first British loss of the war], for instance, would not have been able to penetrate the armor system of the [newly recommissioned, World War II–battleship USS] *New Jersey*. Numerous similar instances occurred in World War II, such as when the Battleship *South Dakota* sustained 45 hits from 8-inch guns [in the 1942 battle for Guadalcanal] and continued to operate, or when the *Musashi* absorbed 14 torpedoes and 22 large bombs and continued to steam ahead.*

No doubt modern conventional weapons can be built with special penetration characteristics that improve on the weapons of World War II. But the report speaks with a voice that should be heard. More ships like the *New Jersey* that can take hits and continue to fight back will complicate any enemy's calculations and compel him

*Secretary of the navy, *South Atlantic Conflict,* p. 3.

to rethink both his weapon characteristics and his tactics in conventional war. Missiles cannot be expended against the warship the way gunfire has been: there are not enough of them.

The modern American navy must be able to adapt to two environments. The ability to sustain damage in conventional war and continue fighting can and should be built into warships, and last-ditch point-defense weapons in each ship should be an integral element of fleet counterforce. The ability to sustain damage in nuclear war and continue fighting is, if not nonexistent, at least much less. Point defenses have little value because of their short range. In nuclear war tactical cover and deception are the primary means of counterforce. A properly armed battleship exemplifies the best means of warship survival in conventional war, a modern submarine the best in nuclear war. Tactics for the two environments are different, and the fleet must be configured, practiced, and ready for either.

Paradoxically, the smaller the war, the more justified is the ship with built-in staying power. Building more carriers and recommissioning four battleships reflect the U.S. Navy's desire to retain its ability to fight small wars. Nevertheless, the trend is a shift away from hull strength. The recommissioning of the battleships redressed a balance that was tipped too far, too soon, on the side of a trend that nonetheless will probably continue.

We should always remember that survivability is a characteristic incorporated to gain time for the offense. Critics who talk about surface ship vulnerability ignore this. The less knowledgeable among them assume that expensive ships should stand up in combat forever; the wiser among them contend that big ships are not worth the money —and if someday there is an alternative that delivers superior net force, that is to say delivered firepower over a ship's combat lifetime, they will be correct.

Important to understanding these discussions is the way the fleet tactician looks at defensive force. Defensive systems collectively act like a filter (not a wall, or Maginot line) that extracts a certain number of incoming aircraft or missiles. As it is able, a hull absorbs hits and allows the warship to conduct curtailed offensive operations.

World War II AAW weapons destroyed some air attackers and, with a curtain of fire, distracted others. Modern hard- and soft-kill defenses do the same. Up to a point, the defense takes out a high percentage of the attackers. When the attack is dense and well co-ordinated, an active AAW defense will be saturated at a certain point, beyond which most missiles or aircraft will get through. The modern concentrated air or missile attack aims to reach beyond the saturation point of the defense.

Two other trends bear mentioning. One is the growth of a tactical no man's land, a region where neither side may operate its main force and where pickets (aircraft, submarines, and missile craft) will fight fierce subordinate engagements to create weakness or gather information. The no man's land exists because defense needs room. In conventional war, battle space translates into time to react against attack. In nuclear war, it may be that no defense is adaquate and space is needed simply to stay out of reach or to make it too difficult for the enemy to target moving ships. A smaller no man's land has long existed. In the past, daylight surface actions with guns did not occur at less than two thousand yards: action was fatal before the range closed to that point. Battle lines did not expect to fight at ten thousand yards, the zone where destroyers lurked. Carriers did not want to approach other warships closer than one hundred miles—a miscalculation or an adverse wind would put guns within range, and it would all be over in fifteen minutes. HMS *Glorious* discovered this, and many more of the U.S. jeep carriers off Samar might have at the Battle for Leyte Gulf if Kurita had not lost his nerve and retreated with his overwhelmingly superior Japanese surface fleet. Today the zone of danger extends to five hundred nautical miles and farther.

The second trend concerns the vulnerability of ships in port. Ports have traditionally been havens for navies superior and inferior. Though few harbors have ever been absolutely safe from attack, the strategy of the nation with the weaker navy has been heavily influenced by the consideration that a fleet-in-being could be reasonably safeguarded in port. But this has changed; the security of ports has diminished. Pearl Harbor, of course, earmarked the transition, as did several other striking if less well-known events. On the night of 11 November 1940 a handful of torpedo planes from HMS *Illustrious* surprised the heavily protected Italian fleet at the port of Taranto.

They put three of six Italian battleships out of action for six months, one for the rest of the war. And the Italian fleet fled to Naples.* It was not long before Japan also learned firsthand the results of in-port vulnerability. After Sherman's carriers struck Rabaul in November 1943, the Japanese navy was so stunned that it soon withdrew to Truk. Not much later, in 1944, carrier strikes penetrated Truk, and the Japanese, unable to challenge at sea and completely frustrated by the U.S Navy's ability to concentrate overwhelming air power against any island bastion, withdrew into the western Pacific. Today, almost half a century later, ships are often safer outside of home port than in.

Scouting

The goal of scouting is to help get weapons within range and aim them. Scouting gathers information and reports it. The dominant trend in scouting has been the increasing rate of search and the increasing range of reconnaissance, surveillance, and intelligence-gathering systems. The reason is obvious: longer-range weapons demand these improvements. Less obvious is the reason scouting has had to struggle to keep up. Weapons fire in any direction. In figures 6-1 through 6-5 we saw firepower represented in one dimension. Now we will consider an aircraft reconnaissance in World War II whose search area looks like that shown in figure 6-6, a pie shape. Double the range of the enemy's attack aircraft and you quadruple the area to be searched. A barrier search—a scouting line—can sometimes cover the perimeter of this expanded area. The bent-line screen invented late in World War II to detect submarines in front of a carrier is an example. Still, tactical commanders cannot often be satisfied with a scouting *line*. For one thing, it is usually pervious: submarines that can approach submerged and launch missiles are a threat that seemingly springs from anywhere at or inside missile range. For another, searches cannot always be continuous. In World War II, when scouts or patrol planes in tactical support were launched by the Japanese and Americans at dawn after a night without reconnaissance, they were never sure how far out the enemy might be found.

*Roskill, pp. 110–14.

Doubling the range of search (of any given density) requires quadrupling the area of coverage around a point.

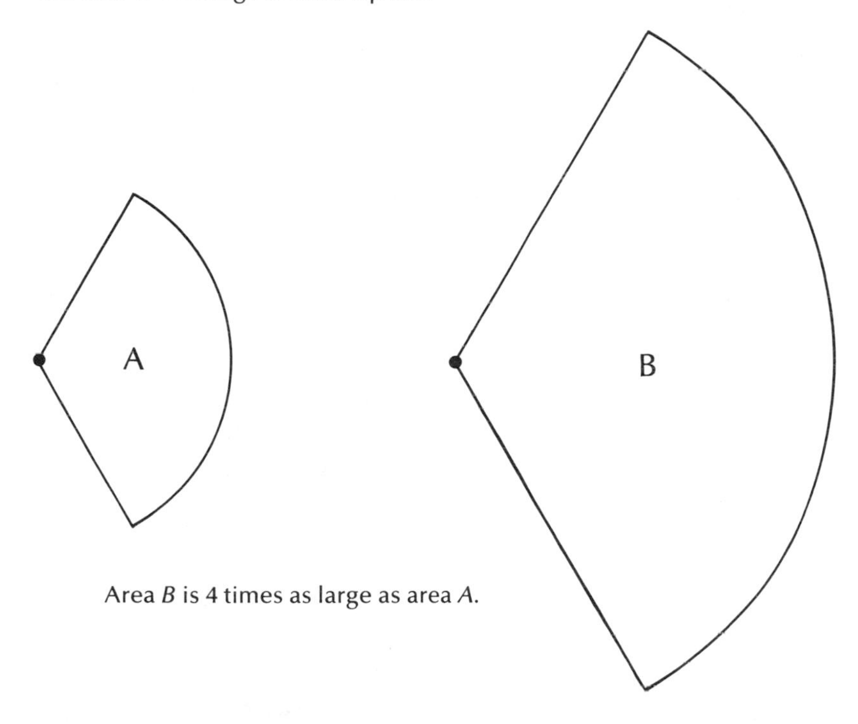

Fig. 6-6. Illustration of Search Capacity as an Area Concept

Most commanders today would like to track every ship and aircraft that might be a threat. As weapon range has increased, the area involved, expanding as the square of weapon range, has stretched surveillance capacity to the limits.*

There is a second development that has affected scouting. Recall the scouting line thrown out in advance of the World War I Grand

*Dr. Joel Lawson cogently argues that, given government's willingness to pay the bill, modern surveillance systems could track and display everything at sea—about seventy-seven thousand naval vessels, merchant ships, and aircraft. He also knows that such a capability is no more than the basis of a tactically useful system. See his chapter in Hwang, p. 63.

Fleet. Its placement was governed not by gunnery range but by a calculation of the time it would take to relay a signal to Jellicoe by wireless plus twenty minutes to allow for the shift from cruising formation to battle line before the enemy closed to weapon range. As we have noted, in warfare space is equivalent to reaction time. Now that missiles can approach at twice the speed of sound or greater, reaction time is so compressed that the scouting line must account for both missile range and the time it takes to act against air, surface, or submarine launch platforms. Some authorities illustrate this with three circles or pie slices. The smallest is the region of *control:* any enemy inside it can and must be destroyed. The next is the region of *influence* or competition, something like a no man's land. The largest is the region of *interest*: an enemy inside it must be prepared against. Scouting in the first region seeks targeting data; in the second, tracking; and in the third, detection. Thus the effect of the increase in weapon-system range and speed has been not simply to increase the area in which weapons may be delivered but to expand the size of the battlefield so that it includes the entire region of scouting and preliminary maneuver.

The vertical dimension of the battlefield has also been extended: deeper beneath the surface and higher above it. This complicates the tactician's thinking enormously. He has to deal simultaneously with operations on three levels, each with its own set of weapon ranges and constraints and scouting capabilities. Ballistic missiles cannot turn, torpedoes have drag, and air-to-air missiles have g-limits. The commander's bedevilment is that tactical decisions must be woven into all three planes. Submarines torpedo enemy surface ships, of course, but the real complication is that when submarines or surface ships fire missiles, a subsurface or surface threat transforms into an "air" problem. The tactical commander is not playing three games of simultaneous chess; he is playing one game on three boards with pieces that may jump from one board to another.

Warfare is in the process of extending into space. Space will be a fourth plane of action, as different tactically from the air as the air is from the surface. No one knows how space combat and tactics will evolve, but we may be certain that space warfare will happen. James Dunnigan writes in *How to Make War* that "air operations revolve around the gathering of information. They always have; they still do."* The first wartime role of aircraft, on land and at sea, was scouting. In this aircraft were so successful that the antiscout—the pursuit plane—was invented. In World War I all other aircraft roles were by comparison inconsequential. The same sequence of events in space is certain. Space satellites are already enormously important for surface surveillance. In some particulars they are peerless for scouting. Like antiaircraft fire in World War I, earth-launched antisatellite systems are not the best countermeasure, nor are the current means of cover and deception. As a direct result of the fundamental importance of scouting, "pursuit" systems in space will be invented to destroy surveillance satellites. Whether space bombers will follow some day we will not guess. But if the past offers insight, it may be that a descendant of Giulio Douhet will prophesy the end of ground combat with the arrival of strategic space bombers, and that a twenty-first-century Billy Mitchell will prematurely predict the victimization of all warships from space.

Antiscouting

Before the age of the big gun, the only cover at sea was incidental gunsmoke or accidental fog. With the advent of the big gun, weapon evasion was recognized as an important tool of war. Surface warships could avoid or delay being hit chiefly by putting up smoke screens, chasing salvos, and combing torpedo wakes.† These were antitargeting devices—ways to confound the enemy's weapon delivery.

As the destructiveness and range of weapons grew, the means of surviving weapon attacks diminished and the emphasis shifted to reducing the enemy's scouting effectiveness. Antiscouting became possible when scouting started to be carried out at long range to

*Dunnigan, p. 98. As creator of good war games and critic of bad tactics, Dunnigan is to this generation what Fred T. Jane was before World War I and Fletcher Pratt was before World War II. If Dunnigan does as well as Jane and Pratt, he will be almost right about seventy-five percent of the time, which is not so bad.

†The idea behind salvo chasing was this: If the enemy observed shell splashes that fell short, he would correct by adding range to the next salvo. If you steered in to close the range, you would do so in the hope that his next salvo would be too long.

account for the phenomenal growth in weapon range. Antiscouting by cover, deception, and evasion would now aim at limiting detection, tracking, or targeting.*

Submarines stayed submerged to avoid detection as well as attack. Carrier task force commanders tried to stand aside, avoid detection, and attack enemy carriers first. Spruance decided to guard the beaches at Saipan and Guam because he thought the enemy might move in around him. The cover of weather was exploited especially by the Japanese, and the cover of night allowed high-speed carriers to run in and attack airfields ashore. Land-based aircraft usually outranged the carriers, but their airfield locations were fixed.

Radar was a terrific scouting system, but passive radar detectors could pick up the enemy at longer, tactically significant, ranges. Radar countermeasures gave early warning but not countertargeting information. For German U-boats trying to evade—not fight—the aircraft of the British Coastal Command, passive detection was enough. For the Japanese in the Solomons early detection was important, but it did not give targeting information, which they needed to deliver firepower.

Communications countermeasures were, by World War II, as important as search and targeting countermeasures. RDF and code breaking positioned the enemy sufficiently to concentrate an attack. To cite another example of the importance of cryptanalysis in World War II, German surface raiders, merchant raiders, and an extensive network of resupply ships and tankers were located and swept from the seas in 1941 not by aircraft scouting, as was commonly believed, but by reading the coded traffic passing to German vessels.†

A natural result of search and communications countermeasures was counter-countermeasures. At Cape Esperance Scott turned off his flagship's radar at great tactical cost; he thought the enemy would

*Other terms are used: stealth for cover; distortion and disinformation for deception; obfuscation for evasion with jamming or decoys. Cover prevents the enemy from knowing your presence. Deception makes him think you're elsewhere and draws his fire away. Evasion ruins or delays his attack. Ervin Kapos likes to speak of "C and D" as confusion and dilution. Another complication: in the U.S. Army cover means protection. Soldiers take cover by digging foxholes or getting behind rocks.

†Beesly, pp. 91–97; Hughes and Costello, pp. 153–55.

otherwise be alerted. The whole structure of communications since World War I has been based on the reasonable proposition that when you talk, the enemy listens. During the war the assumption was usually that plain talk would be understood, while encoded talk would alert the enemy, sometimes giving away one's location but without revealing anything about intentions. As we know now, using code was riskier than this. Also, simple factors such as the time it took to encrypt and decrypt messages and the chance of error or misunderstanding being added during transmission tremendously frustrated commanders when time and accuracy were vital. For at least the first six months of the war, American tacticians manifestly underestimated the friction and confusion caused by communications security.

People who think that automated encryption and decryption have solved the problem would do well to reconsider. Beyond the threat of pilfered information, the equipment costs money, there is never enough of it, and incompatible interfaces turn up at just the wrong moment, especially in joint operations such as the invasion of Grenada. These and the other possibilities inherent in very-long-range warfare portend, on the face of it, antiscouting opportunities and scouting inhibitions in combinations almost beyond our grasp. The trends of warfare say this will be so.

C^2 and C^2CM

In C^2, although the trends have less importance than the constants, there are observable changes. One is the rise in the tactical uncertainties of impending battle. A modern tactical commander works mostly with electronic clues about his enemy (often even about his own forces). A commander of fighting sail saw much more of battle than today's commander can see through his electronic eyes. But still he could not see everything. Jellicoe and Scheer, Spruance and Nagumo, Tanaka and Burke, all had crucial information denied to them and had to make pivotal tactical decisions while operating literally or figuratively in the dark. If this lack of information were simply a tactical constant and no more of a burden than before, it would be sobering enough. But a reasonable conclusion from naval history is that the next battlefield will spring more surprises. Now, without good scouting, the enemy's missiles can come any time, and with such speed as was never possible under sail and only sometimes possible in World War II. A commander at sea faces a twenty-four-hour war. The nighttime Battle of the Nile was an anomaly in 1798. In modern war night action will be commonplace.

The development toward the increasing value of surprise is directly supported by the research of Barton Whaley. In his book *Stratagem: Deception and Surprise in War,* he analyzed eighty-six land battles that took place from 1914 to 1953. Later he extended his study, covering twenty-five more cases that carried him up to 1973.* Whaley concludes that during these years the use of deception increased in order to achieve surprise, and that, moreover, surprise became harder to achieve without deception. For similar reasons, deception probably grew more important on the oceans. Today new and remarkable surveillance and reconnaissance systems make it more difficult for a fleet to achieve a coordinated attack without giving the enemy some inkling that it is coming. But the range, complexity, and redundancy of scouting systems and the complicated process of fusing their products opens new doors to the ingenuity of a tactician disposed to deception.

The continuing pressure on tacticians will be unparalleled. What sort of combat leader is suited to this new environment? Men of youth, vigor, and moral and physical stamina, as important as they have always been, will assume even greater prominence.

Another consequence of the twenty-four-hour battlefield is that ships will fight important engagements at condition watches. In World War II it took a while to adapt to condition II and III watches† and dawn-and-dusk general quarters alerts. Husbanding his crew's energy became an around-the-clock concern of every captain in the war zone, especially during the Battle of Okinawa. That battle, with the kamikaze threat and the continuing pressure for days on end, is the

*His extensive research work was completed in 1969. It was supplemented by Whaley and Ronald Sherman in Daniel and Herbig, pp. 177–94. Although strategic deception was Whaley's emphasis, research by Captain William Van Vleet, USA, led to similar conclusions on the value of tactical surprise and deception in land warfare. Van Vleet also records the most common methods of success on the battlefield.

†Readiness conditions less than general quarters—the former kept roughly one-half of the crew at battle stations, the latter one-third.

best laboratory we have had to observe the psychology of the modern battlefield. We know that the pressure on ships' crews was just as remorseless and enervating as it was on captains. War's prolonged tension is a new phenomenon. Crews will suffer fewer hours of boredom and more hours of fear. In today's navy, ship design and manning incorporate more and more the requirements for battle under conditions II and III. In such combat, tactical commanders and their captains will have to imbue a sense of presence-in-absentia, because the action may be over before the captain is at his station, won or lost by an officer who asked himself, "What would the captain do?"

The great deadliness of modern surprise attack can be illustrated numerically. When guns answered guns, a 2:3 disadvantage could not easily be offset by surprise. For example, according to Bradley Fiske's model of exchanged broadsides, to gain equality the inferior force (call it *B*) would have to fire for ten minutes unanswered by *A*. That is about sixty percent of the time it would take *A* to eliminate *B* if both sides exchanged fire. For *B* to obtain a 2:1 advantage over *A* before *A* started to return fire, *B* would have to fire unanswered for twenty minutes, the military equivalent of the *Chesapeake* being caught unprepared by the *Leopard* in 1807. Compare this with the model of carrier warfare in World War II. If *B* with two carrier air wings could surprise *A* with three, *B* would sink two carriers at a blow and have instant superiority. Coordinated modern missiles have the potential of inflicting similar shock on a fleet. A surprise attack of the scale from which a fleet might recover in the age of big guns will be decisive in a modern naval war. The range of weapons has increased the demand for scouting range and the speed of weapons has increased it more.

As the potential for sudden, coordinated shock attack grows, and that is the obvious trend, the roles of C^2 and of countermeasures against the enemy's C^2 take on new and compelling significance. A modern tactical commander will expend relatively less of his energy on planning for and delivering firepower, and relatively more on planning and executing his scouting effort and forestalling that of the enemy with antiscouting and C^2 countermeasures. Why this is so becomes clear when we consider that the modern equivalent of obtaining a gunnery fire-control solution involves what are loosely thought of today as reconnaissance and surveillance systems. I doubt that the delivery of major and decisive surprise attacks on a *fleet* are now possible, barring the special kinds of tactical surprise that have always been possible even when the enemy is in plain view, as at the battles of Quiberon Bay, Trafalgar, and Narvik. But the evident direction of tactical developments will heap more and more hot coals on the heads of fleet commanders and staffs as they try to compensate for deadly surprise with C^2.

7

The Great Constants

Maneuver

One problem of combat theory is how to define the beginning and end of a battle. Does the exchange of lethal force (firepower) open the battle? Consider the story of the cobra and the mongoose, told by Norbert Wiener in his book *Cybernetics.* The mongoose has the peculiar ability by some combination of mental and physical agility to stay ahead of the cobra's capacity to strike. At the right moment the mongoose attacks behind the cobra's head and the fight is settled. Does the battle consist of one leap by the mongoose? No. Nor does combat begin when the first shot is fired. Sun Tzu, Liddell Hart, and John Boyd would insist that combat is more than the application of firepower, a view I share. The battle proper includes antecedent *maneuver* (not strategic mobility) that bears on the outcome. In Mahan's words, tactics is "the art of making good combinations preliminary to the battle as well as during its progress."*

*Mahan, p. 10.

Throughout history the purpose of maneuver has been to establish a superior fighting posture. Fioravanzo gives us a clue to the constant that connects maneuver old and new. He refers to the fundamental tactical position as the relative location that affords earlier or greater concentration of firepower.* Speed and time, which are dynamic, translate into position, which is static. Even in the age of fighting sail, admirals knew the importance of maneuver before ships came within fighting range and lost speed from damage. In the age of the big gun, the greater speed of battleships was more than offset by the speed with which the battle might end from gunfire. Maneuver played its part principally before ships opened fire. We know the situation today: with a potentially huge battlefield and fast-acting weapons, maneuvers of even the most agile ships appear to be carried out at a snail's pace. Still, some superior position is sought. Speed and time are needed for position, and foresight is the contribution of the tactical commander. In sum, the modern commander must not be deceived by distances—what he takes to be strategic movement may be a battlefield maneuver. Nor must he forget that, while position is his aim, speed and time are his means.

In peacetime the wartime advantage of more speed in combatant ships has usually been overrated. High speed is expensive in money, weight, and space. Somehow peacetime planners fail to address the tactical problem of a formation being tied to the slowest ship in the force. The effect of damaged units on the speed of a force was and still is often overlooked in peacetime tactical discussions. When he was in command of the Little Beavers (Destroyer Squadron 23), Arleigh Burke told his squadron that he would never leave a damaged ship, but he admits today that he was speaking with his heart and not his head. The usually astute Fiske was so taken with speed that he placed it first in importance in his 1905 Naval Institute Prize Essay, ahead of both "manageability" (C^2) and firepower. Mahan, however, was not deceived; he spoke of "homogeneous speed" and influenced the decision to build pre–World War I battleships with arms and armor at the expense of speed. Baudry scornfully referred to armored

*Fioravanzo, p. 209.

cruisers not as the analogue of cavalry, which compared with infantry was "an arm *deluxe,*" but rather as an engine of war "on the cheap." "Whoever heard of squadrons of zebras," he wrote, "ridden by children armed with sticks?"* Jackie Fisher, father of the battle cruiser, a ship whose fatal tendency it was to blow up under the briefest of fire, would have done well to heed Baudry's contempt for speed. Winston Churchill wisely preferred fast but well-armored battleships that could operate ahead of the battle fleet in support of the scouting line.

Modern naval analysts have been notably unsuccessful in making a case for the cost effectiveness of speed, and to the extent that they have succeeded it is mostly in relation to defensive, not offensive, tactical maneuver. Neither the hydrofoil nor the surface-effects ship has proved its case; the speed of these vessels brings too many penalties in its wake. Even the new faster attack submarines were rationalized on sandy ground.†

Frank Uhlig, publisher of the *Naval War College Review,* points out that carriers must be swift to operate aircraft. It is fascinating to speculate what their speed ought to be if this were not so. Since the widespread use of VSTOL aircraft in the future is possible, the question is not idle. The cost penalty of vertical lift (which is the VSTOL's greatest liability) could be offset substantially by reducing the propulsive power of the *whole formation.* We should remember that with half the propulsive power a ship can go about eighty percent as fast. Additionally, speed creates noise in the water, and noise draws submarine missiles. There are times when the ability of a carrier to operate its aircraft at low speed—or even at anchor—is a valuable attribute. But I demur. Strategic speed, for example when a force moves into the Indian Ocean, remains a precious capability. And

*Baudry, p. 47.

†Speed in solo performers such as submarines and single aircraft conducting low-altitude penetrations is subject to its own analysis. Questions of the homogeneous speed of a force and what to do with damaged ships become moot. The thing to bear in mind is that by definition solo performances are unconcentrated, and without concentration a unit depends for long-term survival on remaining undiscovered. And speed, unfortunately, is usually the enemy of stealth.

speed for tactical evasion and countertargeting is at least of some importance.

In World War II the *North Carolina, Alabama,* and *Iowa* classes were the only battleships that could keep pace with the carriers, and it seems pointless to ask if another class of AAW ship might have played the fast battleships' role better. When new technology offers more speed without much compensatory cost, we should embrace it. But when technology offers speed only if this or that is given up, we should not be beguiled. History as well as analysis tells us that that extra bit of speed in ships and speed and maneuverability in aircraft are dearly purchased.

Firepower

At sea the essence of *tactical* success has been the first application of effective offensive force. If the tactician's weighty weapons substantially outrange the enemy's, then his aim is to stand outside effective enemy range and bring down his attack with sufficient concentration of force to destroy the enemy. If the enemy outranges him, then the tactician's aim is to survive any blows with sufficient residual firepower to carry out his mission. Against nuclear attack, U.S. naval tactics have centered on using cover and deception long enough to bring first carrier air power and later SSBN firepower into action. Against conventional Soviet naval air, surface, and submarine attacks, U.S. battle force tactics call for the concentration of enough firepower to defeat an enemy first attack; these tactics depend on net offensive and defensive superiority to dominate, even when the U.S. force is outranged by long-range bombers or forced to take a first attack at the war's onset. As the margin of U.S. naval superiority has narrowed, tactical difficulties have expanded.

It is all the more important now for a U.S. tactical commander to have the means to concentrate effective firepower and deliver enough of it to accomplish his mission before the enemy can bring decisive firepower to bear. If he does not have such means, he should not wish to engage the enemy, for he is likely to lose with very little to show in damage to the enemy. The second great constant of offensive force applies here: Other things being equal, a small advantage in *net* combat power will be decisive and the effect will be cumulative.

The necessary margin of superiority, however, widens when the enemy for any reason can be expected to deliver a first, but inconclusive, attack. The inferior force cannot assume a defensive position and exact a substantial toll as in ground combat. An inferior fleet must be disposed to risk and find a way to attack effectively first. Otherwise it should be ordered to avoid battle and adopt a strategy of evasion, survival, and erosion, which it must hope to achieve with good fortune and skill. The modestly inferior force will usually lose, and with little compensation.

In the previous chapters we have seen some of the theoretical and empirical bases for this conclusion. It was also buttressed by some of the strategic gaming at the Naval War College before World War II. There the results of engagements between "detached squadrons" were decided by the following sort of set-piece evaluations: If there were two forces with relative strengths (not simply numbers of ships) of 2:1, the inferior force was removed from the game. When the odds were 3:2 the lesser force lost one-half its strength. With odds of 4:3 the superior force defeated its adversary but was incapable of carrying out any large operations during the remainder of the game.*

My own numerical estimate is that superiority in net combat power of 4:3 has been conclusive at sea, except in the case of an effective enemy first attack. An advantage of 3:2 will crush the enemy. At times nations have sought a numerical advantage of 5:3 or 2:1, but those numbers are based on strategic rather than tactical considerations.†

Another recurring tendency, perhaps common enough to be called a constant, is to overestimate the effectiveness of weapons before a war. The abysmal ineffectiveness of naval gunfire in the Spanish-American War came as a shock. By 1915, after ships' fire-control problems had been largely straightened out, ten or twenty minutes

*McHugh, pp. 4-28 and 4-29.

†In negotiating the Washington Treaty, the United States, on the basis of strategic responsibilities in both the Pacific and the Atlantic, held out for a 5:3 advantage over Japan in capital ships. At the beginning of this century Great Britain, in fear of a two-power alliance against her, had a two-power standard, which meant her navy was to be as large as the next two strongest navies combined.

of accurate gunfire was conclusive. Nevertheless, at Jutland the High Seas Fleet escaped destruction because the British battle line was unwieldy, the German fleet maneuvered skillfully, and smoke obscured the scene of action. Before the Pacific carrier battles commanders were too sanguine about the effectiveness of air power. And the chaotic night surface actions did not at all reproduce the clean, decisive battles played out in prewar board games because firepower was not as effective as expected. This rule abides: Watch for the fog of war and do not underestimate the propensity of the enemy to survive your weapons. In the next war at sea we will see ships with empty missile magazines and little to show for the expenditure of what some regard as the decisive weapon. When Admiral Burke, the last of our World War II tacticians, was asked what he would change in the new class of guided-missile destroyers, his namesake the USS *Arleigh Burke* class, he said he would add a brace of cutlasses.

Still, the possibility of a decisive outcome has almost always been at least latent. We have observed times when offensive effectiveness was confounded by bad tactics (under the permanent fighting instructions) and by a combination of good armor and poor shooting and maneuvering (after the Battle of Lissa). But in general offensive firepower has dominated the defense, and we should be no more surprised by the destruction of the *General Belgrano* and the *Sheffield* in the Falklands campaign than Beatty was when two of his battle cruisers blew up in five minutes. Nor should we be surprised that HMS *Hood* proved to be too delicate a greyhound; that the Imperial Japanese Navy's back was broken in one morning at Midway; that the U.S. Navy with overwhelming naval superiority was nevertheless losing more than one ship a day to kamikazes in the bloody Okinawa campaign. In modern battle, ships and aircraft will be lost at an agonizing rate. But we observe no trend toward *greater* destructiveness; we see a continuation of naval combat's decisive and destructive nature.

Is there inconsistency between one paragraph that says offensive weapon performance will be overestimated and another that says naval combat will be bloody and decisive? The reconciliation lies in this: even though tacticians will need more offense than they foresee, the offensive capacity for great destructiveness and potential decisiveness will still exist. Dewey and Sampson won decisive battles with horrible gunnery. Jutland may have been tactically inconclusive, but the battles of Coronel and the Falkland islands were certainly not. Even the inconclusive naval battles in World War I hung on a knife's edge, minutes away from decisiveness. Carrier air power was decisive enough to sweep the Pacific Ocean almost clean of carriers after four big battles in 1942. That airplanes were less effective than predicted was important, as it influenced their tactical employment; to the Pacific theater strategists this hardly mattered.

Counterforce

While the success of defense against firepower has waxed and waned and at present is on the wane, the importance of diluting or destroying enemy offensive firepower continues. We may summarize these characteristic constants of counterforce:

— Except for brief periods and in unusual circumstances, defense at sea never dominated offense in the sense that Clausewitz and other observers of land combat intended to convey. The potential for decisive attack at sea has almost always been latent.*

— Defensive force has demonstrated unanticipated resiliency. Its contribution is seldom greater than to give time to attack effectively. But if too much is not demanded of defense, new means can be found to impede new threats.

— Defense will sometimes look more effective than it did during planning because offense will not be as effective as it is calculated to be in peacetime.

— Offense and defense will both fumble at war's onset. But offense will show better early in the war.

*Clausewitz (1976) wrote, "We maintain unequivocally that the form of warfare we call defense not only offers greater probability of victory than attack, but that its victories can attain the same proportions and results" (p. 392). He was speaking of the defensive *battle,* and this in a tactical sense, so he is being taken in a fair context. It is well to add, however, that elsewhere, and in a broader strategic context, Clausewitz wrote, "If the defense is the stronger form of war, yet has a negative object, it follows that it should be used only so long as weakness compels, and be abandoned as soon as we are strong enough to pursue a positive object" (p. 358).

Another constant of maritime warfare is that navies are difficult to replace. For this reason ships of the line did not engage forts with the same number of guns, battleships did not venture into minable waters, and aircraft carriers did not attack airfields that based similar numbers of aircraft. Ships did attempt such actions if they had preponderant force in the sea-shore battle and if they had established sea control. The Gallipoli operations in the spring of 1915 illustrate both the prerequisite preponderance of force and the hazards of engaging shore batteries in minable waters. Three French and British battleships and a British battle cruiser were sunk or damaged, and the fleet's attempt to penetrate the Dardanelles was called off on the very brink of success.

Carrier operations against airfields bear similar risks. Compared with damaged aircraft carriers, damaged airfields may be reconstituted quickly. In conventional war, there is less possibility of concealment, survivability, and recuperation at sea than on land. The compensatory virtues of warships have been their greater mobility and potential for concentration. Nuclear war alters these generalities. Because of the capacity for both strategic movement away from the threat and tactical movement out from under a missile attack, surface warships will be more durable than land-based forces. The capacity for survival of submarine-based SLBMs through concealment exceeds that of land-based ICBMs. Nuclear war also changes the replacement equation: conventional naval forces are more difficult to replace than conventional land-based systems, but in nuclear war, when no warheads are replaceable, this liability disappears altogether.

Scouting

Sun Tzu wrote:

> Now the reason the enlightened prince and the wise general conquer the enemy whenever they move and their achievements surpass those of ordinary men is foreknowledge. What is called 'foreknowledge' cannot be elicited from spirits, nor from the gods, nor by analogy with past events, nor from calculations. It must be obtained from men who know the enemy situation.*

*Sun Tzu, pp. 144–45.

This is found in the chapter entitled "Employment of Secret Agents." We can imagine Sun Tzu fairly rubbing his hands together on his first encounter with modern cryptanalysis and surveillance satellites.

The naval commander has always sought effective scouting at a range consistent with his weapon range. That is, he has sought data about enemy forces far enough away (remembering the time-movement relationship, we read "soon enough") to deploy for effective offensive and defensive action. And his data has included a plot of his own forces. An amateur who imagines a chessboard war cannot conceive of the frustrations of keeping this plot. It is not rare in peacetime exercises for a commander to target his own forces. Every professional should reacquaint himself with the hazard and reread, for starters, Morison's detailed accounts of the Solomons night actions, including the Battle of Cape Esperance, in full and sobering detail. The choice of tactics must be compatible with force proficiency. Unpracticed, widely dispersed forces on a modern battlefield dense with long-range missiles run great risk of self-destruction.

The great constant of scouting seems to be that there is never enough of it. In the days of sail a line of frigates was thrown out ahead to perform strategic search (in those days the great naval problem was to find an enemy at sea at all). There were few other means of knowing the enemy's strategic objective or where he was operating. Like modern satellites or OTH radars, sailing frigates doubled as tactical scouts, and there was no conscious distinction between this role and the strategic one. When contact was made, the frigate scurried back within flag-signaling range. Then the ships of the line had ample time to form in column. Without enough frigates, fleets under sail could be caught in disarray. The Battle of Cape St. Vincent is one such embarrassing instance among several for the French and Spanish in the Napoleonic Wars. Naval commanders cried out in frustration for more frigates. There never seemed to be enough.

In chapter 3 we saw the tremendous amount of force involved in scouting. Jellicoe committed twenty-five percent of his heavy firepower, and Scheer nearly as much. By World War I scouting resources involved more than ships and aircraft. Both sides also tried to exploit signal intelligence. (Scouting by my definition includes the delivery—but not the analysis—of the information.) It is hard to see how fleets without the wireless would have managed their scouting problem; radio communications, handy for fleet maneuvers, were vital to successful scouting.

Between the world wars aircraft became the principal scouts, for detection, tracking, and—as gunfire spotters—targeting. The French admiral Raymond de Belot, writing of the World War II battles for the Mediterranean, said that the Italian navy was continually pleading with the air force for scouting aircraft and was demoralized without them.* The U.S. and Japanese employed submarines and land- and sea-based patrol planes for early warning or strategic scouting. The U.S. carrier force added large numbers of dual-purpose scout bombers for tactical reconnaissance, and to great effect. The Japanese, slower to use carrier aircraft for search, sought to rely more heavily on cruiser-based floatplanes and, whenever they could, land-based search. Their skimping on reconnaissance aircraft cost them dearly. The best commanders never hesitated to augment single-purpose scouts with dual-purpose ships and aircraft, sacrificing fighting mass in order to find and target the enemy. The contribution to U.S. tactical effectiveness of air search radar engaged in detection and targeting is probably undervalued in the history books. Despite the misuse of radar in the early Solomons night actions, the benefits of superior air search and radar search on the Pacific battlefields can hardly be overstated.

Tactical commanders today may also believe they will never have enough for search efforts. The distribution of all search equipment according to range, center bearing, and spread-angle will be one of their most important tactical decisions. As Spruance, Mitscher, and Halsey did with their scout bombers and Jellicoe did with his fast battleships, commanders today will sometimes have to sacrifice massed firepower to augment scouting. If there is global nuclear war involving ICBMs and SLBMs, tactical scouting will play its role as it has in other wars. Resources in space and on the ground will be necessary for targeting, assessing damage, and estimating residual fighting capacity. What will technology produce as the next "radar" device for

*Belot, pp. 41–44, 67–68, 71–72, and 86.

that war, and will we use it well or squander it, as we did in the Solomons with outmoded tactical plans?

It seems pedestrian to say that scouting has always been an important constant of war. Perhaps the way to put it is this: winners have outscouted the enemy in detection, in tracking, and in targeting. At sea better scouting—more than maneuver, as much as weapon range, and oftentimes as much as anything else—has determined who would attack not merely effectively, but who would attack decisively first.

C^2 and C^2CM

Dr. J. S. Lawson, who until he retired was the chief scientist of the old Naval Electronics Systems Command, uses the term command-control to mean the process by which a commander exercises authority over and directs his forces to accomplish his mission. Around 1977 Lawson and Professor Paul Moose of the Naval Postgraduate School devised the decision cycle depicted in figure 7-1.* Their system entails decision (part of commanding), sensing (part of scouting), and acting (largely a part of control through communications), as well as all tactical processes. By incorporating "sensing" and "acting" into the C^2 *process* and building them into a feedback loop, Lawson and Moose and their compatriots allow C^2 to look outside of itself. They also make possible the study of its effect on the activities of firepower delivery and scouting—for example, on the rate or quantity of firepower delivered, or on the rate or quality (tactical significance) of reconnaissance and surveillance information. With the command-control cycle it is possible to examine the tactical context and "productivity" of C^2, observing the steps involved in the delivery of firepower, in scouting, in C^2, or in all three. A tactical analyst, should he choose, may hold two processes constant while exploring the third in detail. For example, he might assume some tactical scouting and

*Independently, Dr. Geoffrey Coyle of Shape Technical Center established a similar paradigm. There is also unclassified evidence in Soviet scientific research in cybernetics that the Russians produced the equivalent C^2 model as early as the 1960s. See Abchuk et al., and Ivanov et al., which are revisions of earlier works published in 1964 and 1971. James Taylor, pp. 36–41, has a useful commentary on this aspect of Soviet research on C^2.

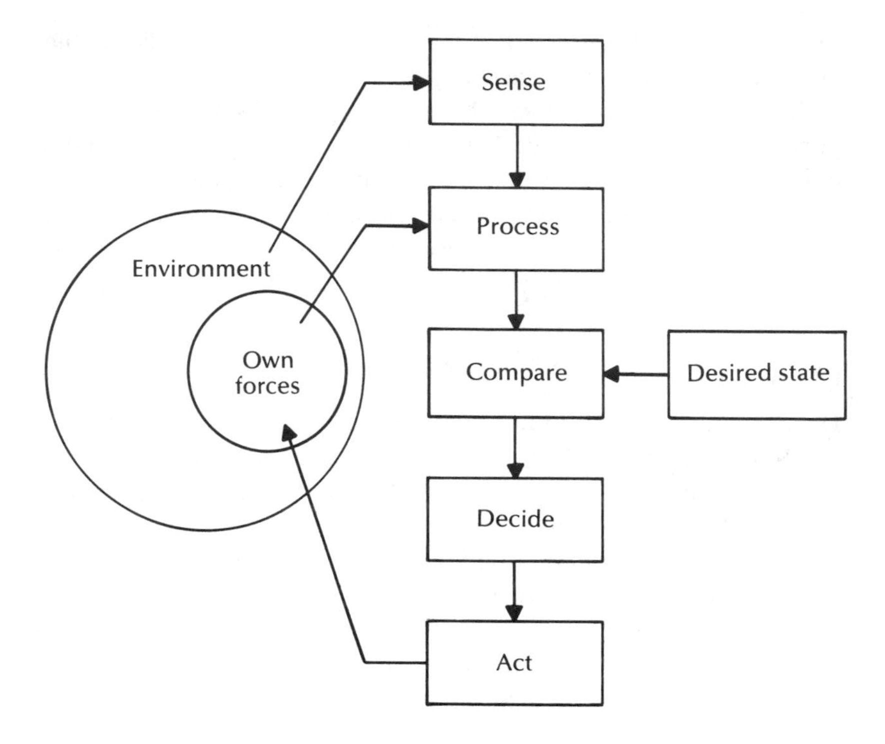

Fig. 7-1. The Lawson Command-Control Cycle

weapon delivery plans on both sides, and then devote his attention to the movement of necessary information into and within the flag center (that is, to the command process). Alternatively, he might study the flow of orders to units on communications circuits, for example to localize and attack a submarine, with emphasis on traffic routing and the time involved in consummating an attack (the control process).

One of the flagrant deficiencies of Lawson's 1977 model was that it originally treated control as a one-sided process. Lawson now believes his model should accommodate an enemy control cycle, which would operate on both the battle environment and one's own forces.* The resulting force-on-force activity is depicted in figure

*From conversations with Lawson in Monterey, California, winter 1985.

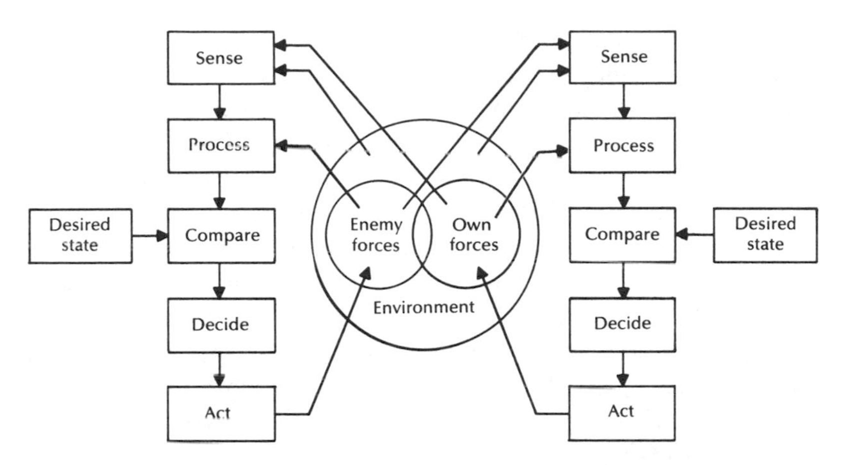

Fig. 7-2. Enemy and Friendly Command-Control Cycles Operating Simultaneously in the Same Environment

7-2. The picture seems obvious, but there is nothing obvious about the complex analytical problems that result from the simple step of including the enemy's control cycle. To begin with, we might like to know the effect of operating, say, twenty percent faster inside our control loop than the enemy is able to do. Does this give us a negligible tactical advantage in controlling the battlefield environment, a twenty percent advantage, or virtually total control? I presume that there is no general answer, but the details of the specific tactical problem at hand still must be addressed. Not everyone who has played with the cycle has grasped the fundamental distinction between an engineering loop with feedback, which controls "nature," and a pair of military control loops, which operate with contrary objectives.

Lawson emphasizes the fact that command-control is a process, that is, how to do something. His control model is an abstraction operating in time without specific tactical content. My emphasis is on command, that is, what is to be done. Command is concerned with the distribution of force—the allocation of combat power. Power distribution is the business of C^2. The distribution of force by command is spatial and temporal, of course, and it is also functional (force is distributed between scouting and striking, or between offensive and defensive firepower, for instance) as well as organizational (force is distributed laterally between the AAW and ASW commanders, hierarchically among the electronic warfare staff officers).

Tactical content is closely akin to Lawson's "desired state," the result command wants to achieve. Early in a battle the desired state has much to do with the assignment of tasks and force positions, and with influencing the activities of the enemy through reflexive control, as Soviet military science calls it. Later in a battle, C²'s desired state is the first delivery of firepower to the enemy in effective batches.

C² planning must deal with tactical content (the desired state) in generalities, unless doctrine is fairly well drawn and the nature of naval operations is fairly predictable. Let us explore this important point. One style of ground attack is to direct operations along a front by specifying for each force element a geographical objective, its desired state. Reinforcements are then sent to the places where operations are experiencing the greatest difficulty. Success is viewed as the simultaneous attainment of all objectives. In this case, victory depends on the absence of exposed flanks. A second, contrary style is to strengthen places along the front where operations are succeeding, reinforcing success with the object of snowballing it. In this case, victory depends on a breakthrough followed by exploitation. In naval operations, the successful defense of a battle group depends on the timely augmentation of AAW, ASUW (antisurface warfare), or ASW forces when necessary to handle air, surface, or subsurface attacks. In strike operations one virtue of naval mobility is the threat of attack and the exploitation of enemy vulnerability wherever it is found.

To develop a C² system, including command responsibilities, staff activities, and the hardware and software to support them, the tactical content of operations must be envisioned in more detail. From the outset the difficulty of agreeing on tactical goals and the style of effective command has plagued the necessarily detailed design of systems for a navy tactical flag command center. Where there is no agreement, the alternative is to design the command center and all other C² support in the absence of tactical content; by default tactics will be dictated by C² support-system designs and locations.

A commander and his staff synthesize information, using decision support systems when they will help do the job better. Today modern displays, geographic and alphanumeric, assist in this process. So does artificial intelligence, which emulates the thinking process and (when it surpasses that process) automatically makes decisions. I do not have an example of a military command decision aid that unequivocally decides better than the human mind. But there are many that do part of the job better. Some weapon fire-control systems assign priorities to threats, lay guns, and fire missiles without human modification, and they have existed since World War II. At least one AAW missile system, while still subject to human intervention and override, is designed to operate on preprogrammed tactical doctrine.

Control is the act of executing decisions that have been made. Verbal, visual, and electronic communications are the great instruments of control. Effective planning and training lay the groundwork for control. One measure of the effectiveness of C^2 is the length of time a plan endures before it has to be changed. Another is the amount of communication required to change the plan.

Time and timing are the crucial elements. Time is duration, timing the point in time when the commander takes a proverbial deep breath and issues his tactical commands. Time is a quantity to save; timing is a moment of choice.

Time is saved when compact signals and communications nets are used to process and disseminate orders more quickly. Time is also saved by decision aids. The other great timesaver is teamwork, achieved through training and doctrine. The time killers are lethargy, befuddlement, physical exhaustion, and disintegrating morale. I suspect that many more disastrous tactical decisions have been made by leaders whose spirit was used up and by fighters who were exhausted than the history books tell us. The debacle at Savo Island, the worst battle in U.S. naval history, can be attributed without doubt to exhaustion. It explains the hasty, maladroit American force distribution and the failure of the American picket destroyers to detect the passage of Mikasa's cruiser force.

Timing is the kernel of successful fighting. The ancient Greeks had an expression for that fateful moment of opportunity: Καιρον λαβειν, *kairon lavien*, or the favorable time to seize an opportunity. A tactician constantly wrestles with the question of whether he has enough structured information to launch a successful attack, knowing the enemy is gathering information for the same purpose. And even if the tactician has complete information, that alone cannot assure success between evenly matched forces. Consider chess. Both players know everything and are in perfect control of their forces. The winner is not determined by tactical information. Good timing is the product not only of information but also of native ability and experience.

In other respects chess is a poor simile for war. Its problems are wholly intellectual. Absent are the fog, the threat, the struggle to reduce mental chaos, and the pressure of timing the attack. Burke has been quoted many times as saying that the difference between a good leader and a bad one is about ten seconds. A commander should always have both time and timing in mind. Another difference between good and bad leaders lies in what they see when they look, what they hear when they are told, and what they communicate when they speak.

The first aim of command is to keep control, or as Bainbridge-Hoff wrote, "to keep confusion away as long as possible."* If doing so is only a beginning, nevertheless it is the place to start tactical planning. Recall that Blake and the other generals at sea first set out to stamp order on their heterogeneous forces by forming columns. After order came concentration of force. I have ventured to say that the Americans fought in column during the early night actions in the Solomons partly because they could not expect much more of their ships without losing control. Even later, when Burke conceived his tactic of hitting successively in small teams, two teams were the most he could coordinate. At Empress Augusta Bay, Tip Merrill's cruisers formed a point of reference while two teams of destroyers sprang from each end of the column. The use of three formations resulted in a grand melee and only modest damage to the enemy. Burke's forces separated and for more than an hour he milled around gathering them up. Everyone "lost control."

With characteristic pith Major General Jasper A. Welch, USAF (Ret.), once indicated his criteria for a "perfect C^3I system." They are listed in order of importance:

— Preserving the order and cohesiveness of one's own forces.

*Bainbridge-Hoff, p. 86.

— Controlling the pace of battle and avoiding fatal blunders. (These first two criteria are Welch's "prerequisites to avoid defeat.")
— Ensuring "non-zero effectiveness." (This is Welch's first prerequisite for winning.)
— Optimizing allocations, strategies, or force compositions—C^2 of assured efficiency. (This, Welch says, would rank close to seventeenth in its relative order of importance.)*

Lieutenant General John Cushman, USA (Ret.), says something along the same lines with regard to communications.

> Imagine a meter for measuring the 'Commander's Satisfaction' with his communications, reading from zero to 100, zero being that he had none, and 100 being perfect. If a commander had only very rudimentary communications, barely adequate but at least something, he might be up to 50 or so on the meter. He gets only double that with a perfect system.†

Tactical complexity is a peacetime disease. After the transition from peace to war, a marked simplification of battle tactics occurs. The tactical theorist underestimates the difficulty of executing complex operations in the heat of battle, and military historians are too quick to point out opportunities that could never have been exploited. Even peacetime naval leaders fall victim to this tendency. Cleverness, ingenuity, and complex maneuvers work best for solo performers like submarines and small units that can be highly trained. Extraordinary evolutions, such as the High Seas Fleet's proficiency at executing a 180-degree turn in the midst of battle, must be doctrinal, heavily practiced, and few in number. Since the enemy can be expected to know about anything that has been practiced very much, complex fleet tactics must work even when the enemy is aware of them. Because of the tactical trend toward more capable communications and decision aids, the temptation to equate complex tools with complex tactics will be almost irresistible. Complexity, however, should only be added after great deliberation and much training.

The art of concentrating offensive and defensive power being complicated, it is easy to exaggerate the potential of the enemy to master

*Hwang et al., pp. 4–6.
†Cushman, p. 6–111.

it. While it may be fatal to underestimate the enemy, it is not sufficient simply to plan for the worst and act accordingly. Sometimes we are taught to base actions on enemy capabilities. That is wrong. Actions should reflect full consideration of enemy capabilities, a vastly different notion. Readers who understand game theory know that it is like the commander's formal "estimate of the situation," in which values are attached to each choice-pair. One arrives at the game theory solution upon discovering the best action to take against the enemy's best action, all choices by both sides having been considered. But as with most formal schemes for optimization, the estimate of the situation (with or without numbers) becomes distorted in practice. For one thing, the estimate is static and battle is dynamic: time and timing will add and subtract alternatives. (In the Shenandoah Valley campaign of May–June 1862, Stonewall Jackson *created* alternatives simply by starting earlier and marching faster than the enemy.) For another, new knowledge unfolds unforeseen contingencies. A battle plan must make provisions for, but an estimate cannot predict, ways that the balance of power may shift before the main strike. Consider the Battle of the Philippine Sea, in which the Japanese long-range carrier air strike was doomed before it was launched, because Spruance had already destroyed the land-based Japanese air component and his fighters dominated the airfields on Saipan and Guam, where the carrier aircraft had to land.

The limits of the estimate are seen most fully if we consider the inferior force commander. For him, worst-case planning is not a winning option. One way to use the estimate is to find the enemy option most adverse to you and try to eliminate it, by such means as scouting, deception, or a quick thrust. For the superior commander, a riskless battle means the loss of great opportunities. If he accumulates force to cover every eventuality, he has cheated his neighbor of force or himself out of time, and time is the strategic equivalent of tactical force. Halsey, who fought with more heart than head, was forgiven his blunders because his command was always moving forward in either strategy (when he was commander, southwest Pacific) or tactics (when he was commander, Third Fleet). Nevertheless, when we rank the big-league tactical commanders of history, Halsey trails Spruance, who fought with both heart and head.

The role of the estimate of the situation is a constant. The estimate is as important as ever in its influence on tactical decision, but it never governs a decision. For the inferior force commander, the estimate is important because it offers clues about how to take risks, which during battle are inevitable for the inferior force.

We often hear that the greater speed of communications and the capacity for control of remote forces are great trends. This may be true for the strategist. But a fleet commander in battle would be wiser to think of the speed of decision and the reliability of executing decision as constants. In his valuable book, *Command in War*, Martin van Creveld traces the history of ground combat command from ancient times to the present. Naturally enough he points to uncertainty as the central problem all command systems have to deal with. But, he writes, "the most important conclusion of this study may be that there does not exist, nor has there ever existed, a technological determinism that governs the method to be selected for coping with uncertainty." From smoke signal to telegraph, from radio to communications satellite, technology is a snare that will trap an unwary military organization. Instead of patterning actions after what available technology can do, we should, as van Creveld eloquently concludes, "understand what it [technology] cannot do and then proceed to find a way to do it [the mission] nevertheless."*

Isn't the time it took to read and act on tactical signals by flag hoist or signal light about equal to the time it takes to do the same today between widely dispersed ships and aircraft? Do we recall how the smoke of battle interfered with the ability to read signals in the days of sail? Or the abominable frustration of VHF radios which failed because of a bad patch between the radio shack and the bridge, or of an ASW aircraft which burned holes in the air for lack of a UHF crystal or the right communications plan? We had better remember today's dropped NTDS links as well as the enemy's ability to garble our conversations with jamming. Submarines are still awkward to talk to. Weapon range and lethality open out formations, and open formations must depend on satellite communications, which are not always reliable. It comes down to this: despite advances in the speed and volume of communications, the capacity to command and control a force in battle has not changed. Why does this matter?

*Van Creveld, pp. 268 and 274–75.

Because it affects the timing of the decision and the distribution of orders. Good plans faltered in the past when commanders underestimated the time it took to communicate and overestimated the clarity of their communications. Among the universe of examples, the struggles of Dan Callaghan to manage his unwieldy thirteen-ship column at the Battle for Guadalcanal come to mind, and Scott's confused maneuvers and orders to commence firing at the Battle of Cape Esperance. Technology is hard pressed to keep up with the need for systems to control forces. The modern commander can pick up inestimable insights from history regarding the drumbeat of battle and his own difficulty—there is no peacetime counterpart—in marching to its cadence.

To stay with the cadence a commander and his staff lay their plans. Planning is a function that unites disparate forces into a task force. To achieve the full cooperation of all components, one staff only, the task force commander's, should do all operational planning. The distribution of forces to ASW, AAW, ASUW, and strike operations is a function he must lead rather than arbitrate. The planning and execution of a united scouting effort, including the management of all electronic emissions in his forces, rest on his shoulders throughout the operation and should not be delegated. This does not obviate the decentralized application of force by his AAW, ASW, and ASUW commanders, but in my opinion the commander and his staff must be completely competent in all aspects of fleet operations if the fleet is going to execute its mission well. Since the days of battleships, cruisers, and destroyers, tactical commanders have directed all components for united action. Only the tactical commander can act as the centripetal force.

The development of doctrine is also part of C^2. Good tactics in wartime derive from good tactical study in peacetime. Sound practice comes from sound doctrine, sound doctrine from sound tactical thought, and sound thought from a foundation of tactical theory. That there were few tactical surprises at sea in World War I may be attributed to the outpouring of naval tactical writing in the early twentieth century. Surprisingly, tactical and doctrinal readiness at war's outbreak seems to have little relationship to the pace of technological change or the number of wartime opportunities to observe warships in action. In World War II the U.S. Navy did better adopting naval aircraft for battle than adapting the more familiar gunship to night action. Tactical effectiveness correlates best with the quality of tactical thought in various publications. U.S. naval warfare publications should compare in tightness, focus, and readership with the old fleet tactical publications that preceded them. Articles on tactics should dominate the Naval Institute *Proceedings*, as they did in the period from 1900 to 1910. The hard core of the Naval War College curriculum should be naval operations, as it was in the 1930s. War games should stress not merely training and experience but the lessons learned from each game's outcome, as in the 1920s and 1930s. In intellectual vigor our modern tactical writing should compare with the best in the world. I believe there is a renaissance in tactical thought under way, started by Admiral Thomas Hayward when he was chief of naval operations. But the quality of tactical literature suggests that we are only beginning. I maintain that the quality of our writings is the best clue—better even than the conduct of a peacetime exercise at sea against an Orange enemy—to our probable tactical competence at the outset of two-sided naval warfare.

What, in conclusion, is the greatest of all constants of peacetime command, nothing barred? I am going to indulge myself. My candidate, like nothing else in this book, is a constant for which I can muster no evidence. Army General Cushman, who by contrast has plenty of evidence for his candidate, believes that since in wartime accomplishing your mission comes first, in peacetime preparation to do so should come first.* Perhaps that is right. Should we quote Clausewitz's "Habit breeds that priceless quality, calm" and say that habit, without combat, must derive from training, which takes priority above all else? Good habit is a worthy candidate. Or shall we say, Know your forces, know your enemy, and know yourself? This is all splendid advice, but I would argue that nothing takes precedence over the peacetime commander's job of finding combat leaders. Let him do his best to find them, send them to sea, and keep them at sea, longer than the U.S. Navy does now. Let the first aim of every seagoing commander be to find two officers better than himself and help in every way to prepare them for war. That done, everything else will follow.

*Cushman, p. 4–13.

Summary of Tactical Trends and Constants

Trends in the History of Naval Battles

— Speed in the weapon (torpedo, aircraft, missile) has increased faster than in the platform (ship or element ashore) delivering it.
— Speed in the platform has become subordinate to speed of weapon delivery. Speed of delivery is governed by scouting and C^2 processes as well as the sheer velocity of weapons.
— Maneuverability has become less important for concentrating force and striking first in battle but is still as important or has become more so for evading enemy weapons.
— The absolute and effective ranges of weapons have increased. Effective weapon range has come to dominate mere weight of firepower.
— The lethality of weapons has increased, especially against the naval machines of war.
— Weapon range and lethality have increased the size of the no man's land between the fleets. Scouts and screens occupy the intervening space.
— The trend in defense has been away from staying power (the ability to absorb hits) and toward defensive force (firepower and soft kill).
— Ship architects have overshot the rate of the above change. Staying power should still be built into all major combatants.
— There is a trend toward spreading forces out while using C^2 to concentrate firepower from dispersed formations and dispositions.
— Ships in port and aircraft on the ground have become more vulnerable to attack. Harbors are not the havens they once were.
— Scouting systems have had to race to keep up with weapon range.
— The region requiring scouting has expanded vertically. The effects of air, surface, and subsurface scouts are increasingly interrelated because more new weapons cross the boundaries of the three domains.
— Scouting from space will lead to battles in space.
— The potential for increasing the effectiveness of antiscouting has accompanied the trend toward greater range of weapons, sensors, and communications.

Constants in the History of Naval Battles

— Maneuvering in anticipation of combat is, as always, an important part of battle, but today it is carried out on a grand scale.
— The purpose of maneuver is to achieve an advantage in position relative to the enemy. Absolute geographic position, such as the strong point, is far less important at sea than on land. Absolute geographic position may have major *strategic* consequences.
— The advantage of ship maneuverability is overrated in peacetime. Its tactical value must be weighed against what is sacrificed to attain it.
— First application of effective firepower is the foremost tactical aim.
— Under continuous-fire conditions, a small advantage in net effective force is not only decisive but will succeed without compensatory damage being inflicted by the enemy.
— Applying sufficient pulsed power (single large salvos), a considerably inferior force can win the battle with superior scouting or C^2. The firepower of the inferior force must be sufficient.
— Firepower is less effective than anticipated in peacetime. Nevertheless, there is usually the potential for delivery of decisive firepower.
— Thus far in history naval battles have been fought best without a force in tactical reserve.
— The purpose of counterforce, antiscouting, and C^2 countermeasures is to delay the delivery or reduce the effect of enemy firepower until one's own firepower takes effect.
— Defense does not dominate battle at sea and has seldom been more than a temporizing force.
— Defenses have demonstrated a surprising resiliency to respond to and impede new weapon effectiveness.
— Naval forces take longer to replace than army forces. That fact and the predominance of the offense at sea have led to greater reluctance to risk naval than military forces.
— Scouting capacity consistent with weapon range and speed of delivery has always been sought by tactical commanders.
— There is nearly always a shortage of scouting capacity.

Trends in the History of Naval Battles (*cont. from page 196*)

— Antiscouting and its likely exploitation have become major constraints on enemy scouting effectiveness.
— The possibility of tactical surprise has increased. Surprise may come from the enemy, from one's own side, or by chance.
— The consequences of surprise have become more serious, because either side has the potential to deliver firepower in a sudden, effective pulse.
— Deception (that is, defeating firepower, delaying scouts, and duping command) has become more important and now plays a major role in successful fleet tactics.
— The capacity for C^2 has increased, but C^2 has been hard pressed to stay abreast of the demands on it. Modern decision aids used to assimilate scouting information have growing potential to improve the timing of decisions and compress the time it takes to make them.
— Combat has become a twenty-four-hour hazard. In fact, with early positioning at long range so important now, it will be difficult to define the beginning of a modern battle.
— Tactical commanders have had to devote more of their attention to scouting and less to delivery of firepower.
— The trends in warfare are machine-related.

Constants in the History of Naval Battles (*cont. from page 197*)

— Tactical commanders must be prepared to reallocate forces and sacrifice fleet firepower for purposes of scouting and screening.
— Time and timing have together been dominant considerations in C^2 and in the exercise of countermeasures against the enemy.
— The first aim of command is to keep control. Control is the prerequisite to every success in battle.
— The estimate of the situation and its quantitative counterpart, game theory, are indispensible tools of C^2.
— In applying the estimate or any rational planning process, an inferior commander must be disposed to run risks to win a battle.
— The pace at which control of a fleet can be exercised has not changed much through history. Planning, doctrine, and training as well as combat experience help reduce the possibility of a commander and his fleet being overwhelmed by the tempo of battle.
— The development of complex tactics is a peacetime predisposition. After the first battle, tactics are simplified.
— The constants in warfare are man-related.

8

The Trends and Constants of Technology

Technological Pace and Tactical Change

This chapter is a postscript to the subject of the great trends and constants of tactics. Technology is renowned for the way in which it changes tactics: tactical trends develop because of technology, and tactical constants abide in spite of new technology. Of course, technological advancement and its periodic overthrow of tactics themselves comprise a central constant of warfare. The question is whether the increased *rate* of scientific discovery today translates into a military trend. Do we see more than just the continuing influence of technology on tactics? Do we see acceleration of that influence as well?

We need to know whether the faster rate of technological change gives way to more rapid introduction of potentially revolutionary weapons and sensors in order to answer two tactical questions:

— How often will the effect of new technology be great enough that, exploited in a series of battles, it will affect the outcome of a war? That is, what is likely to be the frequency and magnitude of technological opportunity?

— How well will these technological opportunities be seen and acted on when they arise, so that they may actually be exploited in battle?

Trevor Dupuy in his unpublished papers has accumulated evidence that in ground combat the impact of a new weapon on a war's outcome has usually been local and nearly always transitory. He believes that a technological surprise by itself has never won a war on land, but that technology accompanied by a tactical revolution has. Napoleon's tactical use of mobile artillery was revolutionary; the field artillery itself was not new. It is ironic that the Germans exploited tanks so effectively with their Blitzkrieg, for one of their victims, the French, possessed more and better tanks, and another, the British, had invented them. In these instances the new tools, artillery and armor, were no secret at all. In contrast, when tanks were a surprise weapon and first used in substantial numbers by the British at Cambrai in World War I, they had local successes but could not exploit them. Tank technology, it has been argued, was prematurely squandered by the British, before the tactics had matured. There is the first issue. Is it possible in wartime to develop a weapon—along with the tactics and training—in secret and in such numbers that it will serve as a war breaker? Or will the technological impact of the weapon almost inevitably be local and transitory?

Secret Weapons and Wartime Surprise

Because there are fewer big battles at sea, the potential for decision by technological surprise is greater. At least one weapon is comparable in decisiveness to cryptanalysis, which wrought the great increase in Allied scouting effectiveness: the Korean admiral Yi Sun-Sin's *kwi-suns,* his turtle boats, which in 1592 won two decisive battles at Pusan and in the Yellow Sea against the Japanese.

Another secret weapon sprung long after its invention was the Japanese Long Lance torpedo. As late as the summer of 1943, the American navy did not know exactly what the Japanese weapon was or why it had been so effective. The Long Lance had been developed in the early 1930s, and Japanese cruiser and destroyer men had trained to the teeth with it. Scorn for Japanese technology takes much of the blame for America's overconfidence at the start of the Pacific war, which was almost as foolhardy as German and Japanese overconfidence in the immunity of their ciphers.

Then there is the atomic bomb, though it was not specifically a naval weapon and not numerous enough to be regarded as tactical. The bomb was the shocking weapon that administered the coup de grâce to Japan in 1945. It was a well-guarded wartime secret and no doubt caught Japanese leaders by surprise. The science and technology took four years to develop and only two bombs were built. Is it possible to keep the development of an "ultimate weapon" a secret in peacetime? Evidence suggests that it is not possible, at least not in the United States. Many people in this country believe secret weapons are proper public news. With his book *On Strategy,* Colonel Harry Summers has helped revive an awareness that a country cannot fight effectively for very long without accounting for the temper of its people. The message would not have seemed so new or fresh if more journalists had read Mahan, the Soviet Russian laws of war, Clausewitz, or even Sun Tzu. Now, the nature of a free population, its society, and its government will also decide the extent and kind of secret weapon developments that are possible. The reader should pick up a thirty-five-year-old book, Vannevar Bush's *Modern Arms and Free Men.* Bush concludes that the open society with its greater exchange of knowledge outperformed the closed Fascist societies of Germany and Italy in the exploitation of science and engineering during World War II.* Surely, however, to win the technology war one must have either better science or greater secrecy. One cannot concede both to a foe.

In most instances the hoped-for surprise of new weapons in wartime has been muted in some way. Here are examples of weapons, mostly naval, that brought disappointment in World War II:

— Magnetic influence mines. Germany introduced them against shipping in the estuaries of the British Isles. They were effective but, used prematurely, turned out to be vulnerable to countermeasures.
— Magnetic exploders in American torpedoes. Developed before the war, they worked badly and were a great setback to U.S. operations. In a short war American torpedoes would have been an

*Bush, pp. 193–232.

unmitigated disaster. The British and Germans also had early problems with their sophisticated torpedoes.

— Proximity fuzes. For much of the war they were restricted to use over water out of fear that the Germans would recover one and adopt the technology against U.S. strategic bombers.
— Night fighters. Highly effective, these were too few in number to be decisive.
— Submarines. They were powerful in effect, but their role against warships was well recognized before World War I.
— Sonar. This was a crucial response to the submarine, developed in secrecy to neutralize the threat. It was not enough.
— "Window," the strips of aluminum foil used to jam enemy-fighter-direction radars. The Germans had window early in World War II, but they delayed its application until the Allies used it in the bombing of Hamburg in July 1943. Both sides appreciated the fact that window was a doubled-edged tool of war.
— Jet aircraft, V-1 and V-2 missiles, and snorkeling submarines. All arrived too late in the war to have much effect.

The following are reasons new weapons, secret or known, do not always deliver what they promise:

— Production limitations, as with magnetic mines.
— Testing limitations, as with torpedo exploders.
— Great complexity, requiring skilled operators and integration into fleet tactics, as with radar and night fighters.
— Great simplicity, threatening adoption and exploitation by the enemy, as with window.
— The risk of failure after introduction, as with the U.S. magnetic torpedo.
— Exaggerated expectations, as with sonar.
— The penalty for maintaining secrecy during a lengthy period of development, as with Nazi Germany's secret weapons.

We may conclude that decisive technological surprise at sea is hard to achieve in wartime. Thus there is little purpose for the inventing side to hold back a new weapon in wartime until the numbers, tactics, and training exist to make it decisive. Certainly the instinct of wartime leaders is to rush a tool of war into the fray. Perhaps it is wrong to venture an opinion on this subject without more written on it, but I do think their professional instincts are correct. To the other side, the side responding to a new technology, I can only suggest that it adopt an attitude of wartime wariness but not paranoia.

Peacetime Evolutions and Revolutions

The situation in peacetime is a sharp contrast. New weapons technology is discovered in plenty of time for the enemy to react to it before war arrives. (For secret sensors the evidence is shakier.) But will the tactical significance of a new weapon, developed in peacetime, be grasped in wartime? When a new enemy weapon is underestimated, failure to invest money in countermeasures and failure to expend enough tactical thought and training on an adequate response seem equally to blame. Obviously the tactician's province is to do what he can with his own tools. Should he use them for dispersion? Evasion? Preemptive attack? Should he deemphasize the victimized weapon system and emphasize the use of what is left in new or better ways? If he does not see the threat and prepare, then perfectly visible peacetime developments will turn out to be decisive wartime weapons which might as well have been secret. Therein lies the real danger. What are these weapons? Chemical, space, laser, stealth, and mine warfare technologies are all on the horizon. Do we see them?

Now we come to the more subtle issue, the quickening pace of scientific research. Doubtless this acceleration may be called a trend. But there is little evidence that it has been translated into better weapons any faster now than before. We see in this country slower gestation periods for warships, warplanes, and the weapons and sensors of war. In the Anglo-Dutch wars whole fleets were built in a year or two. If the English or Dutch had invented a new naval weapon during those wars, they could have put it to sea in that same space of time. Today in this country it takes fifteen years to conceive a new warship design, marshal support for it, get the appropriations, and build the first prototype. If modern war is decided by the pace of modern technology, then why do we ascribe a useful combat life of thirty years to our warships and half that length of time to our fighters and attack aircraft? Why do the Russians, who seem to be able to deploy new designs faster than the Americans, nevertheless keep obsolete ships and aircraft in inventory for so long? Decisive tactical surprise is not likely to derive from the acceleration of scientific research, at least not in this country. High tech is the enemy of speedy exploitation.

Nevertheless, let us postulate a new weapon of war that has the potential to change the face of battle—to break open the war. The long-range cruise missile and homing mine serve to illustrate. The navy first experimented with submarine-launched cruise missiles in the 1950s and deep-sea homing mines in the late 1960s. The question facing the community that understands technological promise is how to transform the technology into combat reality at sea. The answer is that technology should be introduced by evolution instead of revolution. The evolutionary approach can work in a free society. Yet Americans persist in haggling over and redesigning every ship and aircraft and sensor on paper to the point of exhaustion before they are produced. In the case of the cruise missile, the navy decided to hold up production for thirty years because it couldn't demonstrate an efficient method for targeting the weapon. Develop the weapon and the targeting system will follow! It would have been sad indeed if development of the major-caliber gun had been deferred until the details of its fire control had been worked out by W. S. Sims.

U.S. treaty cruisers are the perfect example of a successful evolutionary approach. The limits on heavy cruisers prescribed by the naval armaments treaties negotiated between the world wars were eight-inch guns in a ten-thousand-ton standard-displacement hull. The design aim under the treaties became, instead of cost effectiveness, "tonnage effectiveness."* Two ships of the *Pensacola* class were "designed to weight." They were called dogs, top-heavy and poorly armored. Even while the *Pensacola*s were under construction, design lessons were being incorporated into the second class, and every couple of years a better class was authorized. The culmination of the series of four cruiser classes, each an improvement, was the magnificent *Astoria* class of seven superb and much-admired ships. In World War II the *Pensacola*s fought side by side with the *Astoria*s, and it would be difficult to say which class performed better in combat. The

*Rear Admiral Stalbo, in the Red Navy's *Morskoy Sbornik,* writes that even now this is the most important criterion of warship construction. See Stalbo, no. 5, p. 25.

Pensacolas may have been prototypes, but they were not merely prototypes. They were imperfect but effective warships. And they showed the way to something better: the real breakthrough came with the heavy-cruiser designs produced in wartime, the *Baltimore, Oregon City,* and *Newport News* classes, fully outfitted with AAW weapons, search and fire-control radars, and ultimately semiautomatic eight-inch guns.

There are many examples of important hidden improvements in the combat capability of a weapon. One is the rifling of gunbarrels. Another is the improvement made to fire-control systems in dreadnoughts. New engines can barely be detected from an aircraft's appearance but can vastly change aircraft performance. Any change in computer reliability or cryptology is invisible, and any scouting system in space is invisible, at least to an amateur observer. Karl Lautenschläger asserts that the most important characteristic of the Soviet *Oscar*-class submarine is not its great size but the likelihood that its missiles are guided by space-based sensors.* Submarines that depend on acoustic stealth are in a continuing duel to operate more quietly than the enemy, and the quieter they become the more "invisible" they are. All of these important alterations were possible because there was a combat ancestor to improve upon.

On Great Transitions

When a technology is revolutionary in potential, as the sail was during the age of oars, steam during the age of sail, and the aircraft carrier during the age of the battleship, the practical matter of exploiting the new opportunity is exceedingly complicated, even after you know where you want to go. The objective of an orderly phase-in is not to evade the attention of the guardians of the status quo. The Old Guard sees real threats while they are still wisps of smoke on the horizon and many threats that do not even exist. No, the objective is to solve a monstrous transition problem. If new tactics for new systems are difficult to develop, tactics that blend the old and the new are even more difficult. (Consider the false starts that were made when the

*Lautenschläger, p. 57. A former naval air intelligence officer, Dr. Lautenschläger is now on the research staff of the Los Alamos National Laboratory.

new roles for carriers and battleships in mutual tactical support had to be worked out.) Building a new navy from scratch, as the United States did with the New Navy of 1881–1914, is simplicity itself compared with the job of transforming a navy that already exists and plays a vital defense role, as the U.S. Navy does today.

Assuming the technical development of an astounding new weapon system can take place in an orderly fashion, there is still everything else that follows: retraining, integrating old tactics with new, handling logistic support, and attending to a host of mundane details. These are the substantive problems. Arranging such transitions is not the technologists' responsibility, but it is *the* problem with revolutionary technology. Moreover, it is not likely that a less expensive new weapon with greater capability can save much in the defense budget. First, during its introduction, the costs of new production, and of temporarily duplicated training and logistic support, will add to the budget. Second, the advantage to us of a new weapon will soon be an advantage to the enemy, and on what basis do we suppose that he will not budget the same on defense as before and buy more units than we?

But the central technology problem remains how to make the transition from an old navy to a new one. It is a huge problem, as the situation in America today shows. There are two contending schools of U.S. naval thought. On one side are the American aircraft carrier proponents (let it be noted that the Red Navy is starting to build its own carriers), on the other side is almost everyone else in Washington. Critics say the navy should "get its act together" and present a plan to restructure itself in some new, imaginative way. But it is not that easy. Former Under Secretary of the Navy R. James Woolsey once complained there are more views on the correct composition of the U.S. Navy than the navy has warships. Admiral Isaac Kidd is supposed to have said, sometime during his incumbency as chief of naval material, "Our navy is being nibbled to exhaustion by a flock of ducks. The bites aren't fatal, but you spend all your time fending them off." Too many people in Washington can say no to an innovation and no one can say yes. When Vannevar Bush said that the unity of decision under a totalitarian regime was a recipe for making colossal technological mistakes, whereas there was relative efficiency in the prevalent confusion of decision making in a democracy, he could not have anticipated the tortuous system of procrastination that characterizes modern American defense procurement.* Still and all, frustrated tacticians should understand that even in better times there were great burdens on the men who had to choose where defense dollars would go. The noise of modern Washington's quarrels is like squawks compared with the roars of the old-time lions. In his writings Bernard Brodie reminds us of the difficulties that confront men in the best of worlds:

> Men who have been condemned out of hand as unimaginative or unprogressive may simply have been much more acutely aware of technical difficulties to overcome before a certain invention could be useful than were their more optimistic contemporaries. The mere circumstance that one man was proved wrong in his predictions and another right does not prove that the latter was the more discerning observer . . .
>
> This question is closely related to the whole issue of conservatism in high military or naval circles, upon which there has been a good deal of dogmatic writing. It is natural that inventors or cranks should inveigh against the persons or political bodies whom they believe to be capaciously placing obstacles in the way of their own recognition. In the aggregate such condemnation mounts up to such a hue and cry that even the disinterested observer is likely to take it up. Writers vie to surpass each other in hurling invectives at the "big-wigs" or the "brass-hats."
>
> In 1842, Sir Robert Peel, defending the Board of the Admiralty against charges of having ignored Captain Warner's torpedo invention [the whole Warner scheme was commonly referred to in later years as the "Warner hoax"], pointed out some of the problems facing a public body in respect to the adoption of a new invention. "I think," he said, "that on the one hand a public man is culpable if he wholly disregards suggestions of this nature, and on the other, equally culpable if upon slender grounds he lends himself too unreservedly to their support . . . Every man in office has been in the habit of receiving applications of this nature—not a day passes without something of the sort."†

The Story of the *Wampanoag*

Elting E. Morison has been at once a sympathic interpreter of naval conservatism and a critic of it. In one of his essays he tells the story

*Bush, p. 193.
†Brodie, *Machine Age,* pp. 438–39.

of the *Wampanoag,* astonishing product of the technological genius of Benjamin Isherwood which the Old Navy laid up to rot.* On her sea trials in February 1869 she made seventeen knots in heavy seas, and later, returning to New York in calm water, she made twenty-three.† The fastest ship outside of the United States at that time was the British *Adriatic,* which once ran a measured mile in calm water at a speed of fifteen knots. No other ship would match the *Wampanoag*'s speed for twenty years.

The *Wampanoag* was fast because she was designed by Isherwood as a whole ship, with a mission. The mission was to hunt down Confederate raiders, or else, and as an afterthought, to act as a raider herself, preying on British commerce. She was a sprinter. Therein lay the problem. To the westward-looking, insular, Indian-fighting, railroad-building American nation of 1870–90, the *Wampanoag* and in a sense the whole navy seemed to be superfluous. Morison speaks out against a reactionary navy that wanted to follow the old ways and was blind to the future even when it lay tied up in New York Harbor. He is right about the blindness. He is wrong to assign blame to the navy alone; the whole nation was responsible. The grizzled old sea dogs of the secretary of the navy's board who rejected the *Wampanoag* on absurd grounds (Morison lists them) were guilty of being inarticulate—it is an old problem with naval officers—but that is all. They were practical men working with a budget that had been all but eaten up by government apathy, and worse, they were trying to run a navy that was maintaining a distant and poverty-stricken presence around the world. The *Wampanoag* could go like a flash for five hundred miles or cruise for a few days on coal, but what good was that on the African station? With a shallow draft, slender lines, and the ability to move nimbly as a greyhound under steam, under sail she was awkward as a seaman recruit. In the 1870s navy ships had to have endurance to execute their pathetic little mission. It is true that navy leaders might have been more indignant. Instead Vice Admiral David Dixon Porter, hero of the Civil War, champion of sail, and enemy of engineers, would order cruises under sail for economy and to "instruct the young officers of the Navy in the most

*E. E. Morison, *Modern Times,* pp. 98–122.
†Pratt, *Our Navy,* pp. 343–46.

important duties of their profession." While the West was being won, nobody cared about a navy. For most of two decades an unarticulated national policy decreed that there was no place for *Wampanoags*. The decision to lay her up seems sad, but inevitable.*

We are not quite finished with the story of this remarkable ship, so pregnant with technical and military implications. The effectiveness of raiders was on the wane. From the days of Francis Drake to Rafael Semmes and James Waddell, the payoff of *guerre de course* trended downward. In 1875 the smoke of twenty knots could be seen a long way off, and one ship under steam did not evade squadrons of two or three for long. Imagine how a navy of commerce raiders would have affected the New Navy a decade later, around 1889. You can hear the Old Guard complaining: *Fine thing this fancy new doctrine of Mahan's about command of the sea, cultivated by that academic Stephen B. Luce up there in Newport, Rhode Island! We have the fastest cruisers in the world, we have, all descendants of the master-engineer Ben Isherwood's first greyhound! We know our mission—it's commerce raiding. Let's have speed, battle cruisers will be next, then the USS* Indefatigable, *the USS* Hood, *the USS* Repulse. The children of the *Wampanoag* could have been a navy like Mussolini's, fastest in the world, but with never enough fuel to train, hardly enough fuel to go to battle, and not even enough speed to run away.

Effecting a Transition

Everyone should read Elting Morison. He is almost always right about the U.S. Navy. His examples of its technological myopia are withering. He explains with magnificent understanding why this is:

> And the Naval service of that time [mid-nineteenth century] was more than a set of regulated routines, ordered procedures, and prescribed tables of organization. If not quite a whole society, it was, at least, almost a complete culture. . . It can seem at times arbitrary, discriminatory, elitist, insensitive. . . And, like any closed system, it was lim-

*For more on the nadir of the U.S. Navy, also see Albion, pp. 199–204. I am less pro-*Wampanoag* than the historians, who seem blinded by the virtue of technology for its own sake. Until 1902 the British, who also had many remote stations to patrol, built one-thousand-ton sloops of war to cruise under sail.

> iting on the higher flights of imagination and longer reaches of intelligence.
>
> But there were great and redeeming features, some of which, perhaps, it would be nice to have today. . . The whole structure of the Navy was shrewdly designed to enable men to deal effectively with that unforgiving, incalculable element, the sea. Furthermore, it was imaginatively designed to enable men to live and work together in confined spaces, in uninsulated intimacy, for long periods of time in isolation from the rest of the world . . . The meaning of this relationship of the man to the service, to the authority of culture, had in it something of that governing power that moves the churchman.*

In all his writings Morison expresses well why the modern navy needs a research and development base, the vision to see what that base offers, and the power to act on insights. And yet there is the *Wampanoag,* the ship without a mission, to remind us that technological anticipation is not enough. With it go a naval policy compatible with technology, a strategy, a competitive price, and last but not least, an appreciation of the tactical context into which more speed, or more armor, or more endurance, or more firepower, or a better sensor is going to fit.

Usually more than one piece of technology is required to create a revolution. Sail and cannon together replaced the oared galley. Steam power alone was not enough to replace the ship of the line. It took the steam engine, the screw propeller, and the metal hull all together, which in turn made possible the big gun and the marriage of rifling, breech loading, and an effective fire-control system. Big aircraft carriers were nothing without powerful aircraft engines to lift bombloads worthy of the name, and big aircraft required powered elevators, catapults, arresting gear, and the science of long-range navigation over water. The big naval revolutions depended on both a polyglot of technologies and a synthesis of leadership. Even the Polaris submarine, the embodiment of a naval revolution as neat and swift as we are likely to see, would not have arrived without the inspired marriage of two technologies, nuclear propulsion and solid-fuel rocketry; the work of two great technical leaders, Hyman Rick-

*Morison, "Scientific Endeavor," p. 14.

over and Red Raborn; and a chief of naval operations who understood warfare, politics, and the value of swift action, Arleigh Burke.

When the big-deck aircraft carrier that has been the mainstay of American sea power passes from the scene, the change will be grand and intricate—technologically, strategically, fiscally, and tactically. I do not say Elting Morison's truism about navy conservatism does not apply, for indeed even among businesses like the steel and automotive industries, which *do* know the direction they should be taking, there is powerful inertia, a tendency always to defer things for just one more year or two. But in the navy's case there is no clear direction. Besides STOVL and VSTOL aircraft (and all they imply), there are at least three other major technologies that might serve as the basis of a new force structure. When the nature of System X, as I will call it (it will really be a web of systems), becomes clear, the navy must work out its transition problem, for carriers will remain useful instruments of sea power long after System X becomes the sword point of battle. During transitions old technologies linger, and for good reason. This was the case with the battleship—it abided, though it was prematurely maligned after having been almost undone by neglect during the Washington Treaty years. It was the case with small guns, which were still used after the advent of the big gun, whose propitiousness in the beginning was overrated. It was the case with wood and sail when steel and steam were in the offing but not geared to the American mission. It is even now the case with diesel-powered submarines; there were many kinks to work out of the first nuclear submarines, and even today, after thirty years, nuclear submarines have not and should not totally supplant their non-nuclear counterparts in the world's navies.

Strategically, System X must guard the surface and allow the continued movement of commerce and military force across it; and guarding the surface means dominating the air space above it. These are special American national interests, which represent at once both vulnerability and opportunity.

Fiscally, the transition to System X must be endorsed with more understanding than current carrier critics have, whose argument rests on the proposition that a carrier battle group is terribly expensive. Navies have always been expensive. The unspoken reason that drove the Washington Conference to a treaty was that none of the signatories could afford an arms race. Each already knew it would probably have to abandon much of its own construction because of tremendous shipbuilding costs. As I have said, a new navy that replaces 15 aircraft carriers with 150 Systems X is not likely to be cheaper, and the cost of the transition will, or should, exceed "normal" budget levels. The biggest contribution the advocates of a new navy can make is to lay to rest the myth of a cheaper fleet.

Technology, strategy, and budget are all contributions that friendly outsiders may make to the great transition. Fleet tactics for System X will be the navy's own unique contribution. However the system works, naval officers will need to shape it imaginatively, not simply by updating old operational requirements for missile cruisers, fighter aircraft, or communications satellites. The transition from oar to sail was more than a change from line abreast to line ahead; the change was veritably from old tactics like those on land to unprecedented new tactics at sea. Although a column under sail looked like a column under steam, the tactical rationale for one column was vastly different from that for the other. If this book helps lay the groundwork for the new tactics that accompany System X by unfurling some professional sails in the imagination of officers who will someday work them out, it will have served a better purpose than the one intended, which is to spawn a debate over the tactical employment of *current* weapons and the development of tighter doctrine for them.

In any event, System X will not be imposed on the navy from outside. Technologists may offer a menu of alternatives, strategists may propound the policy context, comptrollers and congressional committees may cajole or threaten, the Red Navy may offer new threats to render obsolete old means of sea power, but in the end the navy must find its own way. The aim of this section has been to make clear to the civilian that a major transition is a deep maze. While there is no guarantee that the navy will find its way through the maze in good time, evidence suggests the amateur zealot will only lead the navy into blind alleys.

Summary

Tactical (and, I propose, strategic) change wrought by technology is a great constant. There does not seem to be an acceleration in the advent of new and revolutionary weapons to correspond with the accelerating pace of technology itself.

Occasionally the introduction of a remarkable new weapon developed in the midst of war has had a decided effect on the outcome of a campaign. Usually the effect is limited because of the need for secrecy, testing, complexity, production, and training, and because of the threat of enemy discovery. When a new weapon system is developed during war, it should be put into action quickly.

New weapons and scouting systems developed overtly in peacetime can win wars when they are accompanied by sound tactics and suitable doctrine and used by well-trained forces. Weapons and scouting systems developed carefully but secretly in peacetime can be expected to have important consequences, but these are limited to the extent that production, doctrine, and training are curtailed by secrecy. To respond to clandestine and not so clandestine enemy instruments of war, doctrine must be adaptable at war's outset. Someone outside of the navy's regular training and operating establishment must be thinking about the tactical ramifications of surprises, so that new tactics may be swiftly introduced, practiced, and put into action.

New weapons often require the development of new tactics by men of great vision. Both weapons and tactics will be perfected more quickly if a series of similar fighting machines are built, each model following rapidly on the heels of its predecessor. It is impossible to design the perfect weapon for large-scale production and employment without practicing with it; even then, it takes three or four generations of hardware before a weapon realizes its potential. Observers looking for great breakthroughs and ultimate weapons may miss important changes in capability that are not manifest in a weapon's external appearance.

Tactical hopes and technological opportunity are separated by an invisible wall, which is a source of friction and frustration. Naval tacticians have been guilty of trying to fit new capabilities into the tactical framework with hidebound lack of imagination. Inventors have been guilty of advocating new capabilities, like the *Wampanoag*'s speed, that are too fragile, too narrow in their purpose, or too expensive for tactical adaptation.

As for the great transitions in naval warfare, these take longer than expected, not only because of the time required to perfect a new instrument of war and build it in numbers, but also because a shakeout of tactics takes time. While the new manner of warfare shapes up, plans must be formulated for a transition, during which the old and the new both have combat roles to play. These roles are decided by evolutionary tactics, doctrine, and training—that is to say, by the warrior-customers. But the ultimate *impact* of a great transition in the hands of a master tactician may be felt like a bolt from the blue, even when technology has introduced the new weapons in front of our very eyes.

Great transitions require the engineering insight to fuse several scientific potentialities into a dramatically different weapon or sensor, the tactical insight to see how the weapon will change the face of battle, and the executive leadership to pluck the flower of opportunity from the thorns of government. The inspiration for these transitions often comes from outside a navy. The perspiration always comes from within it.

9

The Great Variables

Theory, Planning, and the Battle's Proximity

In his book, *Discussion of Questions in Naval Tactics,* the noble Vice Admiral S. O. Makaroff related a remark Napoleon made to the Russian ambassador to France in 1812: "All of you think that you know war because you have read Jomini. But if war could be learned from his book, would I have allowed it to be published?"

Yes, if theory won battles, theory would be a state secret. But it does not win battles. Even—nay, especially—communist regimes publish military theory openly in the interest of unified purpose. Theory falls short because it cannot predict the variables that decide battle tactics and outcomes. Theory sees trends and constants but not the contexts of time, place, and policy, those determinants of tactics that are unknown in advance of war, those variables in each commander's equation that change from battle to battle, region to region, season to season. Theorists are limited in their power to assist commanders operating in the context of actual warfare. All tactical possibilities can never be abstracted from a study of theory or history.

If that was all, we could omit this chapter. But consider the influences on battle, of which theory is the most general and remote. The next most remote influence is the burden of responsibility on the peacetime commander. It is an axiom that in wartime the mission is every commander's paramount consideration. In peacetime his whole purpose is to prepare to execute his wartime mission. But what is the mission? Modern U.S. naval forces have a diversity of potential objectives, in terms of the scale of war and the location of operations, that is awesome.

Peacetime commanders are the professional ancestors of men who fight. In the navy's inner circles we honor leaders like William Moffett, Joseph Reeves, and William Pratt who helped prepare for, but who were never privileged to lead, our battles. Others, in charity, we forget—men who were devoted to inspections, paperwork, freshly painted hulls, and elegant wardroom appointments. Peacetime commanders do forget that their first responsibility is to keep doctrine current and train to it. Working machinery, full supply bins, and reenlistments matter too, but since they are more tangible than combat readiness they tend to divert attention from it. As for smartness and impressive correspondence, they are indices of energy and intelligence, but in peacetime they come to be mistaken for substance. All this advice is pedestrian—understood and then forgotten. Peace should be a time for remembering our real responsibilities and for renewing tactics and doctrine.

In wartime some variables are quickly known. There is a national aim and a military strategy. There are theaters of action and choices of forces. The war a navy foresees governs the structure of that navy and constrains and shapes its doctrine. If we have built and trained a force against the wrong enemy in the wrong scene of action, there is little that can be done to fix equipment and doctrine at the onset of war. The U.S. Navy today, having been so heavily committed over the years to widely dispersed operations in small detachments around the world with the aim of preventing or containing war, will be ill prepared to operate as a coordinated fleet of many ships in a less contained war. Fleets fight as they have trained. If it has no experience with battle fleets, the U.S. Navy will only fight competently in units the size of battle groups.

More variables are resolved for the tactician as his battle approaches. For the final planning of it he knows his mission and the orders of battle. But the tactical commander is also constrained to work with what he is given. He may do a little training, he may tune doctrine to his own circumstances, and he may infuse his men with his own spirit and style, but only if he is lucky enough to be given the time. Every commander should be troubled by every change of CO and by every transfer of unit. He should be haunted by the memory of the brave but inept pick-up forces in the Solomons battles. A fleet fights on the momentum of two flywheels. One is fleet doctrine; the other is stability in the fighting force. Woe to the fleet sent into battle with neither. Not even Nelson could have triumphed over both handicaps.

The final influence is the proximity of the previous battle. The first battle of a war is like a football game in which both sides have practiced plays without body contact, or a chess player who has practiced only against the computer. The final context of battle cannot be known until battle is fought and the abilities of leaders, men, and weapons are revealed. Insofar as a tactical commander's choices of tactics are concerned—whether they be solid or deceptive, simple or complex, tightly or loosely controlled—who will not agree that only the battle itself tears away the last veil and reveals its final variables? And yet beware the commander who believes he may establish his tactics as he is waiting for battle. That man will never know enough to make a sound tactical decision. Sound theory, peacetime preparation, wartime experience, and the commander's tactical plans together build the possibility of victory.

Missions and Forces

As battle looms, two things come into focus. The first is the mission. The second is the forces involved. Mission and forces should match like a glove on a hand. The U.S. Navy invented "task forces" years ago to coordinate the two. The task force is a marvelous concept: the assembly of the right forces in the right numbers to carry out an assigned task. The ability to assign forces in proper kind and quantity is a subject to which we will return at the end of this chapter.

Everything governing the tactical commander's plan of operations derives from his mission. We will take this up first.

Mission: The Link to Strategy

Knowledge of mission implies knowledge of geography and oceanography and any other physical factors that will affect tactics. These are known, of course, imprecisely. Not so the mission itself, which, handed down by his superior, is the commander's tyrant. With mission comes forces, and sometimes mission comes because of the existence of previously constituted forces.

Strategy determines the forces and objectives of a battle, but that is not the same as saying strategy dominates tactics. Rather than viewing strategy as something that looks down on and governs tactics, I prefer to see it as Clausewitz does: "[E]verything turned on tactical results. . . . *That is why* we think it useful to emphasize that all strategic planning rests on tactical battle or not—this is in all cases the fundamental basis of the decision. Only when one has no need to fear the outcome . . . can one expect results from strategic combinations alone."* Sound strategy depends on a knowledge of all forces and their tactics sufficient to estimate the probabilities of winning. Thus at the Naval War College it will not do to study strategy and offer strategic plans without first studying in detail the forces and tactics on which those plans depend. Strategy and tactics are related like the huntsman and his dog. The hunter is master, but he won't catch foxes if he has bought and trained a birddog.

Still, a tactical objective is always determined by higher authority in a strategic context. As I have pointed out, sea battles support some larger purpose ashore. In theory, we know that a fleet's foremost objective is the destruction of the enemy's fleet in a decisive battle. The reason is explained by the basic premise of naval strategy: destruction of the enemy's fleet opens all doors. In practice, a great battle for command of the sea seldom occurs unless both sides choose to fight. What Clausewitz said of war applies here to decisive naval battle: the decision for war originates not with the aggressor but with the defense, since the ultimate object of the aggressor is not fighting but possession.† Naval history is replete with examples of one side

*Clausewitz, p. 386.
†Ibid., p. 377.

deciding to avoid decisive battle, which helps explain why there have been so few battles at sea.

There was a time when a group of ships was kept in port as a "fleet in being." The reason for this was to prevent the prospect of being defeated without inflicting enough damage on the enemy to deny him the opportunities he could enjoy after winning. A survey of modern naval weapons suggests that keeping a fleet in being is harder than it was, but it is not yet an outmoded strategy in conventional war.

Another possibility, as we saw in the chapters on World War II, is that the inferior force will fight when and only when the other fleet is at a disadvantage in the pursuit of its operational objective. The goal of naval operations is usefully thought of as either sea control or power projection. Sea control aims at protecting the sea lines of communication, but it usually focuses on the destruction of enemy forces that threaten them. Power projection aims at employing sea control, predominantly by air strikes ashore or amphibious landings. Power projection would be clearer if its definition included the safe movement of shipping and the timely military reinforcement and resupply of ground operations, but usually it does not. We saw how in World War II a commander defending a beachhead or attempting to reinforce one was caught with divided objectives, which no statement of primary mission could entirely clear up. Spruance clung tenaciously to his mission and defended the beachheads in the Marianas and was criticized for doing so. Halsey carried out his mission to destroy the enemy fleet so single-mindedly that MacArthur barely escaped a debacle at the Leyte beaches. In the Solomons one side or the other was always hampered by a mission associated with the fighting on the ground. An enemy can exploit the problems of a superior force whose plate is too full. It is axiomatic that strategy never should allow the tactician's plate to overflow.

Now, the point is that strategy cannot always fulfill its obligation. It is an observable—and understandable—phenomenon that, to draw the enemy into decisive battle, a "projection" operation must threaten him. A major projection operation adds weighty responsibility to the tactical commander's burden because of the reach of modern weapons. The tactical commander must have in mind that which is paramount: destruction of the enemy at sea or safeguarding an operation connected with events ashore. We may observe a paradox in the Pacific war. The Japanese navy usually tried to destroy warships first: sea control was their first priority. The American navy, while schooled in the philosophy that the enemy's fleet is the primary objective, stubbornly protected its beachheads, sometimes at great cost. The obvious conclusion, that assuring the success of the projection operations is the wiser course, is not, however, unambiguously endorsed by the facts. If the U.S. Navy had not had its advantages in radar, cryptanalysis, and shipbuilding, the more classical efforts of Japanese strategy might have been less futile.

There were other examples in World War II of successful projection operations, more single-minded in content, that flew in the face of prior sea control. The North African landings took place in November 1942, even though the back of the U-boat threat was not broken until May 1943. The 1942 campaign for the Suez Canal, which the German general Rommel conducted without having reduced Malta or the British submarine threat in the Mediterranean, nearly succeeded. British naval operations in support of Greece and Egypt were carried out with the most tenuous of sea control. When the Germans seized Norway with no command of the sea approaches, the British Admiralty was stunned and Prime Minister Churchill enraged. The capture of Crete from the air was so costly to the German army that such an operation was never repeated; but it succeeded, and the Royal Navy paid dearly in warships lost during the assault.

These operations are unique to World II. One searches in vain for an example, from the centuries spanning Hannibal's campaign and Tsushima, of an overseas operation succeeding on the ground without control of the intervening sea. For the new phenomenon of projection without sea control I lay responsibility at the doors of the submarine and the airplane. A successful overseas operation conducted without the ability to operate on the ocean's surface, either before or during the action, strikes me as anomalous. The initiation of air strike or amphibious operations, and the support of forces on the ground, require use of the sea, and the dual objectives of these operations will continue to plague tactical commanders. This would likely be as true against Soviet naval forces, which are composed principally of submarines and land-based aircraft, as it was against Japanese forces on land and at sea in the Pacific war. A fleet that cannot bring a projection threat against a continental power is no threat. The Red Navy is not likely to come to the U.S. Navy for decisive battle until (a) the American threat is too serious to disregard, (b) the U.S. operational mission gives the Soviets an advantage, or (c) the Red Navy becomes the superior force.

War's Intensity

Another variable associated with mission that cannot be resolved in advance is the scale, or intensity, of the war. At one end of the spectrum is a crisis somewhere in the world that national interests want contained. At the other extreme is unconstrained general nuclear war. In between there are many gradations. One is theater war with conventional weapons.

Containing crises has been the bread-and-butter mission of the American navy for almost forty years. Crises have averaged more than one a year. Some have been deterred (and thus been undervalued). Others have involved brief combat, but since World War II none have seen successful attacks on U.S. warships on the high seas. Dealing with these crises has been one of the continuing successes of U.S. naval operations.

Up the scale of intensity lies active war with conventional weapons. The war in Vietnam is an imperfect example: it might also be categorized as the upper extreme of a war of crisis containment. Except for the incidents in the Tonkin Gulf, all the naval operations—from air strikes, to naval gunfire support, to "market time" (blockade-like) operations, to the South Vietnam riverine wars—were conducted in support of ground operations. Since the enemy posed no maritime threat, the United States could take the movement of shipping and sea control for granted, and it still does so, too readily. It is dangerous to believe that the U.S. Navy will continue to operate from an ocean sanctuary against other powers without losses. In many minds a significant conventional shooting war at sea with the Soviet navy is a real possibility. This prospect, while not cheery, is so much better than the nuclear alternative that American policy bends every effort to make conventional war more rather than less likely than war at the top of the scale.

About general nuclear war, so awful a prospect, Americans have preferred not to speculate beyond the concept of deterrence. Even military men who speak of absolute deterrence as being unrealistic shrink from thinking through the tactics, C^2, and weaponry to be used in nuclear war. America's firsthand experience with vast destruction is limited to the Civil War. A better example of total war is the Punic Wars, which ended with the utter destruction of Carthage. Russians, who have seen more of war's devastation, have established more realistic plans to fight and survive nuclear war than Americans have.

As for the tactician, while he deplores the circumstances of nuclear war, he is left with the responsibility of conducting it. Lately the U.S. Navy seems to have been shouldering that responsibility with more energy. One of the special characteristics of nuclear war is the series of constraints on the tactician's employment of nuclear weapons. A combat leader whose fundamental maxim is to attack effectively first is burdened with excruciating problems of command, sensor, and weapon survivability in a situation calling for no first use and constrained use thereafter. The tactical imperative is to make sure that the tactician, strategist, and defense policy maker do not have three disparate concepts of nuclear battlefield operations.

Throughout history deterrence has been a function of military force at all levels of war. The correct way to think of the array of naval functions (again with the U.S. Marine Corps a prominent naval element) is in relation to the deterrence as well as the conduct of war, from local crisis to total engagement. Sometimes we hear that deterrence is not a mission. That is to remind us that deterrence not backed by the capability and will to fight is hollow. The marine forces in Lebanon in 1983 were a bluff, which was called and which collapsed. Those who would thrust naval forces into an exposed position (and every deterrent force, from Clemenceau's famous "one American soldier" on the Western Front to the elaborate NATO naval exercises in the Norwegian Sea, is such a thrust) should ponder what lies beyond these tokens of resolve. On occasion there will be a test of will and blood spilled. In 1787 Thomas Jefferson said, "The tree of liberty must be refreshed from time to time with the blood of patriots and tyrants. It is its natural manure."

To sce that deterrence is a weighty responsibility that distracts attention from tactical development, one merely has to look at the U.S. Navy's rigorous peacetime operating schedules and the mischief they create when fleet schedulers try to maintain stable task forces to train together. It helps here to appreciate the communist view that nations at peace are in conflict. Communist ideology turns Clausewitz's most famous dictum on its head. It is not that "war is the extension of politics by other means," but that politics is one way to win the continuing conflict that goes on with or without war. Communist ideology would give Western naval forces no rest even if they had no other functions.

A fire department is effective when it puts out house fires before they spread. It does not have to put out a Chicago fire to be regarded as a success. All the debates over the navy's composition have focused on the fleet's capacity to fight a Chicago fire. Most of the fleet's deployments, however, have been to fight the house fires of the world. Therein lies the dilemma of the U.S. Navy and Marine Corps regarding missions, tactics, training, and deployment. Naval forces must aim to deter at all levels, and when that fails, they must aim to contain at the lowest level while continuing to deter at the next level. No naval force, nor any military establishment, ever before faced such a plethora of responsibilities and mission ambiguities.

Of two tactical implications of this problem, the first involves carrier battle groups. U.S. carrier battle groups are deployed around the world and are valuable for their presence. They will keep the peace, unless a hostile nation thinks that they are inadequate to the task or that, for reasons of statecraft, they will not be used, in which case there will be fighting. At the same time, carrier battle groups ought to be exercising as a fleet, comprised of three or more carriers and their consorts, to develop tactical competency and the doctrine necessary to fight a larger-scale conventional war. Also, the navy must not become so attentive to peacekeeping or practicing for a non-nuclear engagement that its quite different employment in nuclear war is forgotten. In general nuclear war every fixed target will be attacked. Naval forces at sea, when wisely positioned, are the nation's most survivable combat elements. SSBNs are obvious assets, but surface ships, which may serve as command ships for very high-level commanders, as survivable weapon magazines, and as a second strike force, are also valuable. The latter missions receive too little attention in the navy, and outside it carrier battle groups are opposed on the fallacious grounds that they will rush into range of the enemy's nuclear weapons at the earliest opportunity.

The second implication of multiple responsibility and mission ambiguities relates to C^2. To conduct sound C^2 planning, it is necessary to distinguish the characteristics of the different levels of warfare. Most American C^2 planning seems to emphasize the first level, crisis or confrontation, in which the purpose is containment with a satisfactory political or strategic resolution. At this level, naval forces operate under strict rules of engagement. A major tactical problem is to apply pressure with visible presence while facing the continuing threat of an enemy surprise attack. In crisis containment visible presence is an asset; in active combat it is a liability.

U.S. crisis situations have been handled under the tight control of the National Command Authority (NCA). In effect, tactical command has been exercised from Washington, with much of the formal chain of command bypassed and sometimes poorly informed. To strengthen this sort of prompt regulation of events, the NCA has made a great effort to build a C^2 system for direct communication worldwide, even reaching fairly low-level units.

A truism of international conflict is that a nation must succeed both militarily and politicially. During a major war the political elements are subordinated: world opinion and international law are at best slighted, at worst flouted. At the crisis level both military and political considerations weigh heavily; circumscribed force is the order of the day. The military tactician thinks in terms of executing his combat mission with minimum losses to his own force. The statesman, on the other hand, thinks in terms of the political objective that precipitated fighting or the threat of it. These military and political objectives come into conflict. The tactical commander in a crisis or confrontation cannot escape friction between military and political aims; the goals of statecraft confine his military plans.

Though military men at the scene of action and in the chain of command bridle at the amount of control exercised from Washington in a crisis, the record of forty years of crises suggests that such rudder orders are likely to continue. Detailed direction of localized transitory military operations, even those involving shooting, has flowed and will probably keep flowing directly from the seat of government to the tactical commander at the scene of action because of their enormous political content. Thus, it would be wise to construct C² doctrine accordingly. The doctrine should (1) lay down a modus vivendi that unites the NCA so that when it plans and directs operations at the scene it speaks with a single voice (the NCA can be a Hydra whose heads speak with inconsistent political motivations); (2) specify that the chain of command be informed of orders going directly to the scene of action; (3) provide for suitable explanations to accompany the assignment of forces, reinforcements, and logistical support from intermediate commanders; and (4) hedge against the possibility of containment failing and a conflict escalating by preparing for the rapid restoration of the chain of command when it becomes necessary.

I find nothing inherently illogical about detailed direct orders from the NCA if the above doctrine is faithfully followed. Command may be efficiently exercised at the highest level that meets two criteria. First, command's span of control must not be exceeded. Combat activity must be localized so that the commander deals with a manageable set of subordinates. Second, pertinent and timely tactical information must be accessible. With the modern means of scouting and communications, a commander sited in a command post remote from the battlefield may have as much or more data than the on-scene commander in a ship. While experience shows how easily the man in a command post can overestimate the quality and timeliness of his picture of the battlefield, it also shows that the on-scene commander can underestimate the strategic and political implications of his tactical decisions.

If a crisis escalates to the second level, theater war, there is a danger, perhaps the preeminent danger, that communications to subordinate commanders at the scene will deteriorate, that the span of control will overwhelm the NCA, and that the tempo of war will demand local initiative and authority. This is one reason to provide doctrine for a short-circuited command. The NCA may not appreciate the need to relax the reins, as in the Vietnam War when there were too many off-the-scene orders. But the greater danger is the inculcation in our tactical commanders of an attitude that will cause them to freeze and wait for orders when initiative is called for. In theater war, pace and timing will be everything, even as communications deteriorate and the fog of war descends.

At the level of general war, conditions will be even more chaotic. Sound defense requires plans, facilities, and doctrine that will provide for recovery and for continuation toward objectives that range from some kind of negotiated termination of the conflict to sheer national survival. It would be absurd, for example, if defense plans depended heavily on the unequivocal supposition that the president and the Joint Chiefs will survive an attack. Even if they do survive, for a time measured in days many combat components may not know who is in command. To cover such a case there should be preplanned actions that forces would take automatically. Another unsound defense measure is to base all surveillance-processing activities ashore. Many people believe that surveillance information should be collated at sites such as Pearl Harbor, then transmitted in near-real time to the tactical commander afloat. The shore facility is more survivable in theater war, but in general war it will be pretargeted, so a mobile afloat facility is preferable. Airborne command posts are survivable, but ingenious provision must be made for their logistic support after a matter of only a few hours. A seaplane command post—a craft that can sit on the water but also move quickly to avoid attack—is a striking way to combine survivability and greater logistic endurance.

Missions and Strategies of Inferior Navies

Theorists are prone to think in terms of two great sea powers in conflict. The policies of continental powers are also worth examination, because those policies eventually affect the tactics of both land powers and sea powers. We saw what happened when Britain's eighteenth-century navy, accustomed to the aggressive Dutch, attempted to use the same tactics against the reluctant French. British tactics, ill-suited to the new enemy, failed. The French spoke of the ultimate objective, the higher purpose of navies, and French governments would not send their fleets into decisive battle. The British, who understood their own navy's role, scorned this policy, regarding it as the rationalization of inferiority. But the French were right—

the ultimate objective of navies is to influence the lives of populations on the land in one way or another—and often their strategy was sound.

When policy results in an inferior navy, the first thing to observe is that a defense by fleet action will fail. What have been the options of the inferior navy? One is to maintain a fleet in being, as the Germans did with their High Seas Fleet after Jutland and the French often did with their sailing navy. But the competence of an inactive navy withers away and over time the superior navy will be able to take successively greater risks to exploit its command of the sea. A second possibility is to try to whittle the enemy down to fair odds in decisive battle. That was the High Seas Fleet's wartime objective before Jutland and the Imperial Japanese Navy's training objective before World War II. The High Seas Fleet developed tactics that emphasized deception and trickery to gain an advantage in battles between small detachments. The peacetime Japanese developed tactics appropriate to inferiority, which from habit they exercised during the wartime period of Japanese superiority.

A third approach, when the ratio of forces gives the smaller fleet a chance, is to catch the enemy with a temporary vulnerability and exploit it to gain command of the sea. The inferior navy cannot base its actions on enemy capabilities but must be risk prone and willing to act on an estimate of enemy intentions. Doubtless that was what Nimitz had in mind when the American fleet was outnumbered before the Battle of Midway. His orders to Fletcher and Spruance were to fight on the basis of calculated risk.* An inferior navy should put unstinting emphasis on superior scouting. Nimitz and his two combat commanders based their battle plans on good intelligence from code work. To attack effectively first, an inferior force must overcome its limitations by some combination of initiative and surprise.

A fourth approach is to establish local superiority, as the Germans

*In a special letter of instruction supplementing the operation order, Nimitz wrote: "In carrying out the task assigned. . . you will be governed by the principle of calculated risk, which you shall interpret to mean the avoidance of exposure of your forces to attack by superior enemy forces without good prospect of inflicting, as a result of such exposure, greater damage on the enemy" (S. E. Morison, vol. 4, p. 84).

did in the Baltic during much of World War II and the Italian navy and air force did at times in the Mediterranean.

The fifth possibility for an inferior navy is simple sea denial. The goal of sea denial is to create a vast no man's land. Why should command of the sea be necessary for a continental power to achieve its purpose on land? Denying the sea to the enemy may suffice. The U-boat campaign against British shipping in two world wars was an unambiguous attempt at sea denial in the service of continental aggrandizement. The British submarine, surface, and air campaign against Rommel's sea line of communication is another, less pure, example. Sea denial, extended long distances at sea by air and submarine attacks, is an obvious Soviet naval strategy that carries great potential weight.

When sea denial fails as a war-winning strategy, it can still contribute to the war effort if it has leverage. As we noted, the U-boat campaign was a success, at least in terms of the cost of the response it required from the Allies. The kamikaze pilots could do nothing about restoring Japanese command of the sea, but they could hope to cripple and delay the U.S. Navy and thereby help to defend the Japanese homeland. Many believe the Soviet navy's major aim, along with safeguarding their SSBNs, is limited sea denial.

This litany of strategic options for the continental power is not quite complete. There is another possibility that Mahan's disciples tend to slight. The continental power may achieve a maritime objective by action on land. Before World War I, one school of opinion held that Britain must avoid commitment of a large army to Europe. Britain's military role should be to protect its colonies and trade in the rest of the world, its policy to maintain a balance of power on the continent. The other school, among whose proponents was British army General Douglas Haig, maintained that standing aloof from the war on land could lose the continent to Germany. Enemy dominance on the continent, opined this group, might not be fatal to Britain in the short run, but it was, to say the least, undesirable. In World War I, army opinion won the day, the British Expeditionary Force was rushed to Belgium, and it suffered grievously.

The advocates of a maritime strategy felt vindicated, and there was a great public outcry against sending British boys into the trenches.

Nevertheless, in World War II a larger expeditionary force was delivered to the continent. Belgium, the Netherlands, and France were quickly overrun in the spring Blitzkrieg of 1940, spawning the miracle at Dunkirk. Whether or not under the circumstances the British should have been committed to the ground need not detain us. We are interested in the difference in the effect on Britain's maritime strategy of her experience in the two world wars. In World War I, when France survived with Britain's help, the German fleet and U-boats had to be based in the North Sea. In World War II, after France was overrun, the U-boats were unleashed from the Bay of Biscay, and had Hitler chosen he could have devastated Allied shipping with aerial attacks from French airfields. The fall of France made the British navy's operations tenuous in the extreme. Events on the ground had drastically altered and nearly crushed Britain's maritime strategy. Sir Francis Bacon's hoary dictum, "He who commands the sea . . . may take as much or little of the war as he will," has to be considered next to Clausewitz's observation that when one takes little of a war, one is in peril of giving the enemy what he seeks. Not only is the influence of sea power slower and less direct than that of ground action, but it is more affected by events ashore than navy men are wont to admit.

Three points concerning tactical missions have been made. First, while it is true that the object of a mission is clarified as the time of action approaches, a chain of preparations for battle constrains the battle commander's choice of tactics. Second, the nature of these preparations is influenced by the tremendous span of missions that must be contemplated and anticipated in peacetime. Last, naval tactical commanders need to pay more careful attention to the interaction of sea forces with events on the ground. There are three good reasons for this. The first is purely tactical: there will be more interaction in the future. The second relates to the nature of America's probable major enemy: his fleet, his organizational structure, and his military philosophy are all land oriented and foreign to us. The third is that we have inherited from Mahan a tendency to focus attention too readily on the big battle. Maintaining the security of the seas is a big task, with or without combat. Contributing to a land campaign directly with combat power is another big task, and in performing it we should not be heedless of the need to control and use the seas themselves. This book, about sea- and shore-based naval forces, the tactics of those forces, and decisive battles, treats an area much neglected. But it does not espouse the big battle as the be-all and end-all of a fleet. Fleets superior and inferior have other, less ostentatious missions and purposes.

Forces

To carry out his mission a commander is assigned forces. To build a cat's cage a carpenter is given lumber and nails. While the carpenter can measure the materials he needs for his task, the tactician cannot. A tactician can't be sure whether his enemy will be a pussycat or a tiger. Only recently, during the Grenada operation, the size of the beast was underestimated. Fortunately the forces of Vice Admiral Joseph Metcalf, the tactical commander, were augmented and he was able to get his job done.

One purpose of strategic scouting, as we saw in chapter 6, is to determine the opposition. The higher commander decides on the basis of the intelligence estimate what forces to allot his tactical commander, having a margin for error whose size is proportionate to possible inaccuracies. That is the concept.

If the strategic commander has an abundance of forces at his disposal, as at Grenada, time and timing come to dominate his calculations. Assembly and deployment of more force always take more time, and time is as precious to strategists as it is to tacticians. At Grenada time was critical, for Metcalf found newly arrived Cubans in the act of building up their defenses as his paratroopers and marines landed. When the tactical commander makes his own estimate of the situation, the forces he has been assigned imply how thoroughly he can hedge against every enemy capability, or contrarily, they suggest the risks his strategic commander expects him to take.

There is another way to look at forces vis-à-vis mission, which is for the strategic commander to tailor the task to an existing force. The advantage of this is coherence of operations. There are many examples. The Third/Fifth Fleet in World War II was a trained, tactically united entity. The old ASW hunter-killer groups of aircraft and destroyer escorts maintained their integrity as they moved from one submarine target to the next. Every marine ground-air-ship team depends heavily on continuity for effective amphibious assault. In these cases time, not force, is the free parameter in operational planning. The measure of the tactical commander's effectiveness is the pace at which he carries out his missions.

The Correlation of Naval Force

The root of effective tactical action is an appreciation that force estimation is a two-sided business and that not all elements of force are found in the orders of battle. The term "correlation of forces" is so concise and expressive that it is a wonder how military men have been able to communicate without it. It took the laconic Sun Tzu forty-eight words to express less meaning:

> Now the elements of the art of war are first, measurements of space; second, estimates of quantities; third, calculations; fourth, comparisons; and fifth, chances of victory. Measurements of space are derived from the ground. Quantities derive from measurement, figures from quantities, comparisons from figures, and victory from comparisons.

It is intriguing to consider what Sun Tzu, intellectual master of the intangibles of warfare, meant by quantities. The intention of Soviet doctrine for conflict in peace and war is to incorporate all quantities in the correlation of forces and means, including geographical and temporal advantages and disadvantages.

What follows is only one of many ways to parse the elements a commander must correlate.

Leadership

Theory has little to say to potential combat commanders about leadership. To a man, they accept Thomas Carlyle's philosophy that "the history of what man has accomplished in this world is at bottom the History of the Great Men who have worked here," and they reject George Orwell's notion that leadership is only doing what is expected of one as a nominal leader—merely a matter of shooting the elephant to please the crowd.

The correlation of forces involves a comparison of leadership. History offers many examples of the value of knowing the enemy commander's habits. Knowing those of Villeneuve and de Ruyter resulted in the use of different tactics. The Japanese could depend on Spruance's conservatism to tie him to the beachheads. Usually, however, naval operations offer fewer opportunities than ground operations for evaluation in action. The personalities of army generals like Montgomery, Eisenhower, Rommel, Guderion, and Patton were well known and their styles carefully drawn in the enemy's mind. Through the years naval officers have had fewer occasions to telegraph their temperaments, notwithstanding such exceptions as Nelson, Jervis, Yamamoto, Halsey, and—what must be the foremost example of a pair of naval antagonists—Hughes of the Royal Navy and Suffren of the French, who fought each other five times in the Indian Ocean in 1782 and 1783.

At sea it is probably the study of the collective style of leadership that more often pays off. In the world wars a few U-boat and air aces were respected for their individual skills and styles, but it was the large set of data from a series of one-on-one or small-unit engagements by nameless participants that constructed a mosaic of tactical predilections. U.S. naval intelligence attempts to paint a picture of Soviet style that can predict wartime operational propensities, and no doubt Soviet intelligence does the same with American style.

A combat leader's appraisal of leadership is objective. The Duke of Wellington surely knew that Napoleon's presence on the battlefield was worth "another forty thousand men"; but Wellington would have been more interested in his own defensive positions on the battlefield at Waterloo in the certain, objective knowledge that Napoleon, because of his style, was sure to attack before Marshal Blücher arrived with his Prussian reinforcements. Abstractions on the eve of battle are worth nothing; the record of prior actions is everything.

The greatest determinants of victory, apart from strong numbers and successful scouting, are the very things every commander will judge most badly: his own attributes and reputation. All good combat leaders are highly competitive; unfortunately, so are most bad ones. Military leaders are by nature aggressive and confident; they are not introspective. There is little a naval commander can do to improve his self-evaluation, nor would wise theory try to change his personality to make that possible. Of course he will try to cultivate his attributes and reputation, but if his seed is bad and his growth stunted, he will not know it. Under the circumstances, the only counsel is this: The untried commander should assume he has average skill and not presume he can overcome numerical disadvantage with talents he may not possess. Self-aggrandizement is the most common personality trap of military leaders. If a commander has talent, it will grow on him. A good reputation may be worth more on the battlefield than good attributes, and a bad reputation will mute even the best attributes. That is why Napoleon sought "lucky" generals. Explicitly, a tactical philosophy that *depends for success* on its leaders outplanning, outwitting, and outmaneuvering a first-class enemy—which is the enemy against whom to procure forces and lay out doctrine in peacetime—is the epitome of a philosophy of fatal overoptimism.

Training and Morale

"We learn how to do things by doing the things we are learning how to do," observed Aristotle in his *Ethics*. The mechanics of training are to be seen not as an end but as the beginning of effective combat. In football, open-field blocking and tackling are creative skills, but they come from rote practice supplemented by competition on the field. Combat training is also two-sided, that is, both mechanical and creative. Morale itself, both on the football field and at sea, is a relative property.

Theory leaves for the day of battle the final estimate of the states of training. But history says that these are, again, a relative evaluation. If commanders are prone to exaggerate their personal virtues, that is counterbalanced by their propensity to undervalue the state of readiness of their own forces relative to the enemy's. Admirals George Dewey and William Sampson won battles with gunnery that was so execrable that General George McClellan himself would have been justified in taking the lot back for more refresher training. In the circumstances, the Spanish-American War admirals knew that the Spanish fleet was even less ready and appreciated the need for quick, decisive action.

We saw how American ships in the Solomons were thrown together helter-skelter. Yet, it does not follow that the command troubles there all stemmed from this deficiency, for at the same time on the Japanese side the redoubtable Raizo Tanaka suffered as grievously and complained as bitterly about the heterogeneous forces that were thrust upon him. The point is not that incoherence is tolerable but simply that a state of training and the morale that accompanies it have to be compared with those of the enemy.

Hardware

The true orders of battle on both sides include a qualitative estimate of the state of equipment. In ships, inoperative equipment has to be dragged into battle along with the good. Most of the gallant USS *Houston*'s wartime life was spent fighting with one of three turrets out of commission. Propulsion plants are an interesting case. When navies depended on massing for concentration of force, a crippled engineering plant in one ship slowed the whole fleet or else the ship had to be left behind. The advantage in speed that Togo had over Zinovi Rozhestvensky was probably doubled by the pathetic state the Russians' propulsion plants were in after their exhausting eighteen-thousand-nautical-mile cruise from the Baltic to the Straits of Tsushima. Nowadays, with the dispersal of modern fleets in combat, it is more difficult to predict the effect on a fleet's composite maneuverability of one or two ships with poor engineering plants. That is yet another variable for the tactical commander to deal with.

All in all, the numbers, quality, and mix of his own forces on the eve of battle are probably the easiest quantitative estimate for a commander to make, and that of the enemy's forces one of the most difficult. A nice problem of great significance is the estimate of scouting resources. This is difficult partly because of the diverse potential sources of information, some of which will not be under the command of either battle commander, and partly because the active use of some sensors will depend on how the battle itself unfolds and on decisions regarding detectable electronic emissions.

Endurance

Combat endurance is governed by the ammunition stocks and fuel capacity of the forces at the scene of action. Tactics as well as strategy are affected by combat endurance. Dewey suspended action at Manila Bay when an erroneous report told him his ships were almost out of ammunition. A navy study in the 1960s concluded that ASW ships escorting military convoys across the Atlantic could easily run out of torpedoes by firing at false contacts, a prognostication corroborated during the Falklands War when the British navy fired hundreds of ASW weapons against an effective Argentine order of battle composed of but one submarine. After the U.S. Navy's study, torpedo magazine capacity was tripled in the ship class involved. *Tactics* must be altered when there is a real danger of running out of weapons and no resupply is likely.

In the October War of 1973 the Israeli navy's fast patrol boats were able to close and sink Egyptian craft that carried missiles with nearly double the range of the Israelis' own by inducing the Egyptians to empty their missile magazines without effect. Both the U.S. and Red navies appear to be reexamining the number of offensive and defensive weapons they carry. But there is an *upper* limit to desirable capacity. For one thing, magazines are hard to armor against enemy missiles. One does not choose to send ammunition ships in harm's way. For another, missiles, which are expensive, cannot be permitted to go down with the ship in large numbers. The research of a student at the U.S. Naval Postgraduate School, working on his thesis on the "correct" balance between area AAW missile systems and point-defense systems, produced an unanticipated side benefit: he was able to examine U.S. and Red warships to determine the relationship between the expected combat life of a ship (in terms of the number of incoming air-to-surface missiles) and the number of AAW missiles still in the magazines when the ship was put out of action. The results are classified, but the study demonstrates that a staff estimate of fleet endurance should be made with due consideration of the *net* missile firepower in relation to the expected combat lifetime of a force, which naturally will suffer losses. Such considerations bedevil all competent staff planning for battle.

It is important to assess fuel capacities in battle planning. The Battle of the Eastern Solomons in August 1942 gave us a memorable example of what happens when this area is neglected. Instead of three carriers, Fletcher had only two in the battle, because he had sent the *Wasp* south to fuel. She missed all the fighting and was sunk soon thereafter by a submarine. Tactical endurance hardly ever enters into amateur force correlations, and, being a distraction, an aggravation, and a great source of friction, it is rarely given the place it deserves. Knowledge of the enemy's endurance deficiencies can lead to a decisive tactical advantage, as Togo demonstrated at Tsushima.

Resilience

The relative accessibility of mobile support ships for resupply of fleet repair facilities and of naval shipyards for handling battle damage weigh in the commander's plans, not so much on the eve of battle but on those occasions when logistics is the most important consideration. In general nuclear war, tactical commanders on both sides might face a battle with literally no possibility of post-action rearmament or repair if their bases are destroyed.

Another tactical concern is the job of reconstituting an effective fighting capability after an attack. Post-attack scouting for damage assessment and a simple, preplanned roll call ought to be included in doctrine and implemented. Plans for post-attack rendezvous to refuel and rearm should be automatic. How to exploit the enemy's own post-battle confusion is also a consideration. The enemy of resiliency is swift pursuit.

As it is reasonable in war to expect damage and temporary suspension of air operations in a carrier, so it is often reasonable to expect quick flight-deck repairs and a return to offensive readiness. A plan to survive during the interim should be part of any American operation order—and not just for carriers but for any key ship, such as an Aegis cruiser, the commander's flagship, or an accompanying fast support ship that has been damaged. These specific and crucial tactical variables must be thought through and communicated to the force some time in advance, because there will not be time to do so when the operation commences. The worst time to plan a way to deal with battle damage is after it occurs.

Tactical Environment

Meteorology, oceanography, the proximity of land, and the fact that coastal waters are subject to mining also greatly influence the tactics used to carry out a mission. I am not sure I have given this subject all the attention it deserves. The vital point is that battle environment greatly affects the conduct of a mission. Environment is the last of our great variables, and theory can offer little practical advice about it. As former Chief of Naval Operations Admiral Thomas Hayward said, "Every body of water requires a different set of tactics."

Payoff: The Synthesis of Force Attributes

Together, *leadership* and *training* describe force *competence*. *Endurance* and *resilience* describe combat support but connote something more as well. The term *hardware* is more comprehensive than *order of battle*, even after all scouting and communications assets are included, because the hardware evaluation is an attempt to balance force and mission and to assess the state of equipment readiness.

Leadership, training, hardware, endurance, and resilience are pieced together by the commander in a mosaic of offensive and defensive force potential. They may also be arrayed objectively by the staff according to a variety of convenient schemes to indicate readiness by individual unit, or by ASW, AAW, ASUW, and strike capacity. These indicators of force competence are calculated for both sides and explicitly or implicitly weighed against each other to produce a net assessment for the mission.

The character of the *tactical environment* is shared by friend and foe. But it is not to be thought of as affecting both sides equally or in the same way. There are day fighters and all-weather fighters, there are searchers and there are evaders (a dramatic example of contrasting influences can be seen in the different ways the sonar environment affects the antisubmarine predator and prey). The best procedure in practice is to derive values (numerical or symbolic) for leadership, training, hardware, endurance, and resilience, next reduce or expand the values according to environmental influences, and finally correlate the two forces.

Particularly with regard to the enemy, qualitative assessment may guide tactics as much as quantitative knowledge. Nelson did not much care how many French and Spanish warships his fleet would face. His tactical plan was based on the knowledge that his force had a great qualitative advantage in the type of ship-on-ship engagement that characterized the age of fighting sail. In the Solomons Americans usually had a better estimate of where and when a night battle would be fought than what they would be fighting. It seems that Japanese tactics in the Solomons were robust, that is, ideally suited for a variety of conditions of surprise and uncertainty, whereas American tactics were not, at least not in the beginning.

The correlation of forces is always mission-specific and undertaken in the context of the tactical plan. Friendly and enemy force relationships influence the estimate of the situation and the plan of action. But the plan of action must be unified and allow variations for contingencies. As the experiences of Nelson, Togo, Jellicoe, and Burke suggest, a good plan is deceptively simple, a remarkable reduction of comprehensive considerations into a cohesive, feasible essence. It accommodates timely modifications, but because it has a momentum to which every change adds drag and the risk of confusion, variations must be few.

Decision aids help battle planning and execution. Some are mere displays of the location and status of forces. Others help position forces and structure scouting plans. Still others facilitate timely actions. Desk-top microcomputers are a boon to decision making. Doctrine can be programmed into modern decision aids. The Soviet armed forces have incorporated doctrine in their computerized devices. So far as evidence shows, the American navy has not consciously acknowledged either the opportunity or the hazard of doing so. The hazard is not unlike the rigor imposed by Britain's permanent fighting instructions, in which doctrine and the system of transmitting commands coalesced. As surely as the world is undergoing a revolution in information processing, so its navies are in the midst of a computing revolution that will overrun the unwary.

10

Modern Fleet Tactics

Missiles and Maxims

After offering their preeminent maxim of tactical success, Attack, the Robisons wrote, in amplification, "Prescribing an approach which enables an action to be opened with full firepower enforces the soundest of all tactical maxims."* Every reader will feel comfortable with this, a solid deduction from historical research, an old shoe indeed, except perhaps that it eschews the notion of keeping a reserve. There are hints, however, from our examination of the air and surface battles in World War II, that something is amiss when the goal of concentrating offensive force is untempered.

To get at what has happened since the time of the Robisons, let us examine a new model of the dynamics of modern combat. I will not repudiate the concept of marshalling firepower—far from it. The wiser course is to go directly to the processes themselves and let the principle of concentration find its own expression.

First we recall the description of an aircraft carrier attack as a

*Robison, p. 896.

phenomenal pulse of striking power with which an air wing could possibly sink several carriers, but in the course of the Pacific war did not. Second we recall the night surface engagements in the Solomons and the shocking multiship destructiveness unleashed by a spread of torpedoes, analogous in power to a missile attack. It is apparent that a modern warship armed with ballistic or cruise missiles and supported with adequate scouting has the capacity to sink several warships, more than its weight of the enemy. Carrier battle groups are criticized on the grounds that modern nuclear missiles will make them vulnerable to an inferior force, an imputation that we have seen is nearly irrelevant. More apt is the less frequently heard criticism of *Trident*-class submarines which, if they could be targeted, would go down with 192 warheads, more than the number of nuclear weapons that would probably be expended to sink them. These huge submarines seem to have been designed on a cost-effective basis; economies of scale drove the concentration in each vessel of twenty-four missiles, each armed with eight MIRV warheads, without regard for the possibility that over their lifetime the submarines might be caught up in force-on-force combat. No doubt a calculation that admitted even a remote possibility of tracking these submarines or attacking them in port or at dispersed harbors would have resulted in the distribution of Trident missiles to more submarines, even though that is a less expedient way to deploy them.*

The most striking illustration of the concentration of warheads in the modern nuclear arsenal is the MX missile, which carries about ten. A natural but unforseen consequence of SALT I, which counted launchers rather than warheads, is that the land-based MX system is considered destabilizing because it offers the enemy an opportunity to destroy many warheads with one in a first strike.

Now, each of these weapon systems, being an indivisible massing of firepower, potentially vulnerable to some kind of successful enemy targeting and first strike, creates a tactical problem. Let us waive the particulars of each case and analyze the situation in more general terms, abandoning all preconceptions and leaving open all questions

*Besides the cost premium of wider distribution, there is the issue of control: more submarines create more control problems and hazards.

of massing, concentration, and the possibility of a reserve component. We will question one maxim in particular, namely, Always use superior force to attack a part of the enemy while forestalling him from doing the same.

A Little Model of Modern Combat

Let us define the characteristics of some small missile ships. For simplicity we will assume that both sides have these characteristics.

— A *shot* is an offensive missile.
— A *good shot* is a shot that will hit an undefended ship.
— *Firepower* (f) is the number of good shots that can be launched by a ship in a salvo, that is, the striking power of one missile ship. We will say that each ship can launch nine good shots from loaded launchers, or $f = 9$.
— *Defensive firepower* (d) is the number of good shots that will be destroyed or averted by a ship's defenses, that is, the defensive force. Assume each ship can defeat the first shot of an incoming attack, or $d = 1$.
— *Hits* (h) are the number of good shots targeted on a ship minus the number of shots defeated. If f/n is the number of good shots distributed equally over n target ships, then $h = f/n - d$.
— *Survivability* (s) is the capacity to absorb hits. Assume the staying power of each ship is one, or $s = 1$.
— *Counterforce* (c) is the total number of shots that can be defeated or survived before a firepower kill, or $c = s + d = 1 + 1 = 2$.

With the above attributes each missile ship has the net firepower (good shots) in one attack to hit and kill three of the enemy:

$$f - 3c = 9 - 6 = 3$$

Three targets will be killed if the circumstances are right, that is, if there is perfect distribution of firepower, simultaneous attack, and good targeting, and all targets are within range. However, it is apparent that a massed force is much more vulnerable to effective attack, because when one ship in a formation is tracked and subject to attack, all the rest are too, and because in this example the defending ships have no capacity for mutual defense but are only able to defend themselves.

If we construct table 10-1 as we did our table of aircraft carrier pulsed power (table 4-1, page 94), but with the dramatic, threefold increase in net striking power conjectured above, then *B,* if he attacks first, will win handily against odds that look impossible in static comparisons.

Implicitly I have described the tactics depicted in figure 10-1a, which has both sides' forces massed. In the example, ships on both sides carry great firepower that is overexposed to enemy surprise attack. So a better tactic would be to spread the missile ships in the hope that not all would be detected and attacked simultaneously (figure 10-1b), or to commit missile ships one at a time in the hope that at least one out of three would get off a first attack (figure 10-1c). We earlier presumed that something like the tactics in figures 10-1b or 10-1c was embodied in the Japanese carrier battle plans in World War II, the Japanese intention being to deliver a highly destructive surprise attack with one undetected force while the other served as bait.

In the circumstances I have set up, the battle will be decided by scouting effectiveness and weapon range. Less obviously, the choice of tactics will also be governed by scouting effectiveness and weapon range. For the sake of discussion, assume that scouting is accomplished entirely by on-board sensors, and that each missile ship has its own independent chance to detect the enemy. If *B* now tries tactic 3, an Indian-file attack, and if his sensor is as good as any of *A*'s, so that he has an equal chance of detecting any one ship first (we omit the complicating possibility of passive targeting, to be taken up later), then *B*'s chance of detecting *A*'s force before his leading ship is detected by one of the enemy sensors is only one in eight. He will

Table 10-1. First-Strike Survivors (A/B)

	Initial Number of Missile Ships (A/B)				
	2/2	3/2	2/1	3/1	4/1
A *attacks first*	2/0	3/0	2/0	3/0	4/0
B *attacks first*	0/2	0/2	0/1	0/1	1/1
A *and* B *attack together*	0/0	0/0	0/0	0/0	1/0

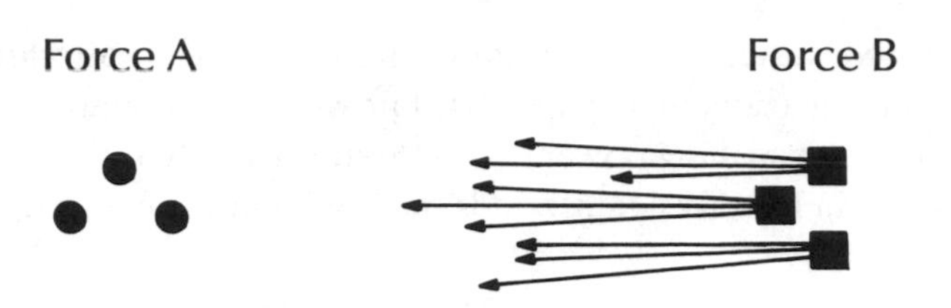

Fig. 10–1a. Massed Attack

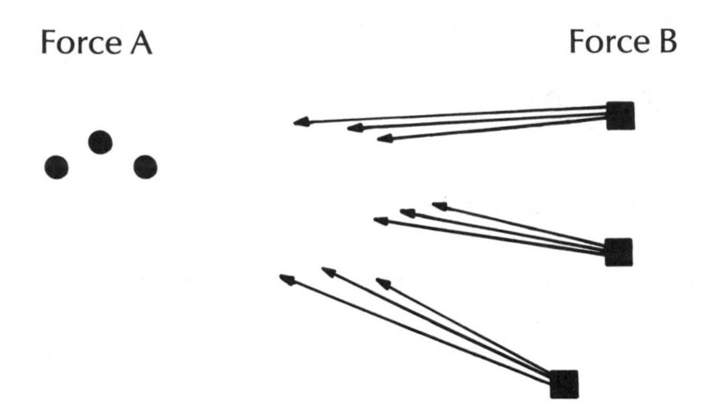

Fig. 10–1b. Dispersed Attack

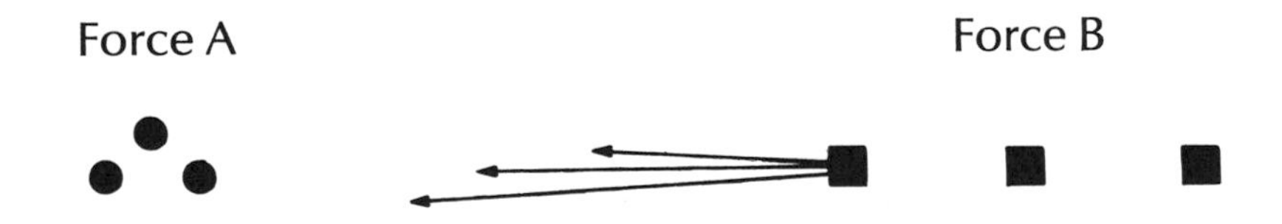

Fig. 10–1c. Sequential Attack

lose the advantage of surprise to *A*'s superior aggregation of scouting rather than to superior firepower.

In the same circumstances, let *B* try tactic 2. In a formal sense, tactic 2 has no advantage over tactic 1. It is more reasonable to assume, however, that if *B*'s units have an equal chance of detecting *A* first, *A* is confronted with the more difficult scouting problem of having to detect all of *B*'s ships *individually* first in order not to suffer the loss of three ships. While it is likely that some of *B*'s ships will be detected first and lost, if only one of his ships detects first, it will devastate all of *A*.

Next, suppose that *A* has longer-range missiles but that *B* has longer-range sensors. As in tactic 2, *B* should be disposed to try and get at least one ship within effective range. If he has good communications, one ship only should be radiating its sensors. Depending on the passive targeting potential of *A*, the radiating ship may be destroyed, but one of the others may be able to close and attack decisively. If one of *B*'s can track *A* while staying outside of *A*'s missile range, *B*'s tracking ship may be able to guide another ship quietly into range with a fire-control solution for a silent attack.

If one or both sides have offboard sensors—satellites, for example—the analysis is quite different. If *A*'s missiles outrange *B*'s, then in the circumstances, the battle is reduced to a contest where *A* alone is stalking. The outranged force *B*, if it must be committed to fight at all, would try tactic 2 and hope that *A* would make a mistake in the coordination and distribution of fire.

Perhaps *A* is predominantly land based. Then sea-based side *B* has a simpler scouting problem against his immobile enemy (precise targeting may be something else). *B* should try to close covertly and attack undetected. But recall the Battle of Midway and the effect American air power located on the island had on Japanese plans: if *A*'s land-based force also has a small mobile sea-based force with potent offensive firepower, then his land base may draw enough of *B*'s attention to allow *A*'s sea-based component to attack with devastating effect.

The above discussion is a paradigm of modern missile warfare. It pertains especially to nuclear warfare, in the sense that offensive firepower per ship is very destructive, the potential to mass defensive firepower in mutual support is difficult, and scouting and weapon ranges favor the offense in new and remarkable ways that encourage the distribution of striking power among smaller ships. Do these circumstances justify Indian-file commitment (or putting it another way, the establishment of a reserve) of forces because of the greater potential destructive power of small units of force against larger units? The answer hinges on the correlation of scouting potentialities.

Ballistic- and cruise-missile submarine deployments mimic many of the attributes of single-file weapon systems: awkward C^2, virtually no defensive firepower, no capacity for mutual defense, and nearly total dependence on first detection and targeting for success.* The force-on-force tactical relationships of nuclear war have not been discussed in unclassified literature. Perhaps open discussion is not yet necessary and may never be particularly desirable. But SSBNs are always subject to search and attack in some restricted way. The ramifications are explored in what will be the definitive work on the subject for several years to come, D. C. Daniel's *ASW and Superpower Strategic Stability*. His book and other less well-developed open studies, however, are concerned with technology, strategy, and policy issues. The tactical side—how the battle would be fought in detail—is at least as significant, for it would be strange and unprecedented, both before and after the exchange of nuclear weapons. I do not pretend to have captured in even a rudimentary way the nature of that battle and its peculiar dynamics. Nor, if I thought I could describe them, would I allow the description to appear in print.

In his detailed technical analysis Daniel shows that SSBNs at sea are very difficult to detect, track, and target, but are easily destroyed upon localization. This is the essence of the tactical situation under discussion. However, it is not only nuclear weapons that alter the classically correct tactics of massing. It is also the suddenness and destructive potential of the modern conventional missile strike. Use superior force is a maxim that by itself is misleading because a markedly smaller force may have adequate net striking power to win. The

*Many imagine nuclear war to be a matter of the pretargeting of fixed sites and the early release of weapons in a general exchange. This is a flawed image on many counts, and in any case our intention is to keep the discussion more general.

dual notions that govern modern tactics are (1) aggregating *enough* force and (2) using scouting and C^2 to strike effectively first with it. As for "forestalling the enemy," the traditional way of doing so, by maneuver or greater weapon range, has to be augmented with modern concepts of antiscouting.

It would be wrong, however, to rush to the conclusion that there is no longer a case for massing force. Let us move on to the circumstances that make massing desirable.

An Illustration of Massing for Defense

In conventional war a primary tactical characteristic is the ability of surface forces to operate in mutual defensive support. The number of offensive weapons is greater, but their lethality is less. Let us slightly redefine our terms as follows, ascribing rough values per major formation, such as a carrier battle group:

— *Striking capacity* (*SC*) is the total number of good shots available in the major formation, including reattacks. $SC = 160$ good shots.

— *Striking power* (*F*) equals Σf, which is the collective firepower of the formation. $F = 40$ good shots per attack.

— *Defensive power (D)* equals Σd, which is the collective defensive force of the formation by both hard and soft kill. The implicit assumption now is that the ships' defending firepower will be aggregated in collective defenses, that is, that area defenses predominate. $D = 25$ good shots averted.

— *Staying power* (*S*) equals Σs, which is the collective survivability of the formation, measured in hits that can be taken. Remaining striking power and defensive power is reduced by each hit. $S = 30$ hits absorbed before a complete firepower kill on the formation.

A single major formation attacked by an identical enemy suffers a net (*F* minus *D*) of fifteen *hits* and loses half of its staying power, leaving a capacity to counterattack with the remaining half of its initial striking power, or twenty good shots. Meanwhile, with his remaining striking capacity the enemy, if undamaged, can mount a reattack of forty more good shots and destroy the major formation unit, whose defensive power is now (formally) curtailed by one-half. Suppose, however, two massed major formations were surprised by a third. If their defensive firepower is combined, then (again, formally) no damage is sustained because the number of incoming missiles (forty) is not sufficient to saturate the defenses with a value of fifty. But if the major formations each operate separately, they are subject to attack and elimination one at a time in tactical detail, because the enemy has ample striking capacity in his magazine reserves.

Even if one assumes that, because of deficiencies in coordinating two major formations, the combined defenses of the two are less than additive (say, the defensive power to avoid only thirty instead of fifty hits), the advantage of mutual *defensive* support is still evident and genuine. For the numbers here, such support is also decisive, allowing the defense to deflect most (all but ten in the formal sense) of an enemy first attack and still counterattack with five-sixths of its firepower, or sixty-six good shots.

There is nothing very surprising in this. It simply shows why the modern decision to mass depends on defensive considerations. When defenses are as strong as in this example, massing the way U.S. carrier fleets did in World War II is not only attractive but mandatory. If, however, defenses are weak, then a dispersed force is indicated. And if C^2 is capable of achieving offensive concentration of firepower over great distances, then tactic 2 (broad-front attack) is required. In certain circumstances related to scouting or to weapon-range advantage or disadvantage, tactic 2 may also be wise, even without the C^2 to coordinate the attack. Nevertheless, the tactician should be aware of the new possibility that tactic 3 (a sequential attack) may be superior in certain situations that are now more likely to arise because of the existence of weapons able to inflict sudden and massive destruction. Essentially, it was such circumstances that governed Burke's tactical one-two punch in the Solomons campaign.

Thus far, for clarity, we have not developed the vital interrelation of active sensor search and passive enemy intercept. It is the most complicated of all considerations, whether the purpose is to aid in developing battle doctrine or to conduct an operation. We have also practically ignored the possibility of very asymmetrical force compositions, such as the U.S. and Red navies have today. We will take up these crucial considerations shortly, when we introduce a final force-on-force model. But first a recapitulation of the significance of the discussion up to this point is in order.

A Recapitulation

A warcraft with great offensive firepower and little means of defense is very vulnerable and creates a highly unstable tactical situation. It depends for effectiveness on a first strike—a stealthy attack or a better scouting–weapon range combination. This sort of craft is a soloist.

A warcraft with a mix of attributes like this is an anomaly. Why is such a "mistake" built? Ostensibly because it is cost-effective to put many good shots (aircraft or missiles) in each craft when the measure of effectiveness is simple firepower, *f,* which ignores the force-on-force nature of battle. A better measure of effectiveness is "deliverable firepower," which is a combination of offensive firepower and counterforce, but it is hard to analyze two aggregated forces that have offensive and defensive properties and are subject to many tactical variables such as geography.

A proper analysis of deliverable firepower would still probably argue for massing offensive power, simply because of the economies of scale and the low cost of adding missiles on the margin. Nevertheless, instability is a consequence of massing, and the above discussion suggests the need for both (1) technological measures (more defensive force, derived from such features as point defenses or chaff, as well as greater hull survivability), and (2) tactical measures (such as plans to attack either Indian-file or from many directions to confound the enemy's scouting and fire control). These measures would give a naval force the temporizing ability to permit its offense to act.

The first question to answer regarding tactical configuration is how much striking power would be sufficient to eliminate the enemy threat at one blow. If your firepower is concentrated to meet this standard of offensive sufficiency, there is no purpose in adding more for effective attack.

The second question concerns the massing of forces, and that is answered by an analysis of defensive capabilities. Together, concentration for the offense and the possibility of massing for the defense determine the tactical disposition. However, these calculations are not made simply by a correlation of weight and range of firepower. C^2 determines whether dispersed firepower can be concentrated offensively. Scouting resources (one's own and the enemy's, on board and off board), range and search plans, and so forth affect both offensive and defensive effectiveness. Antiscouting potential also affects the final correlation.

One of the practical problems is that the correlation of force cannot be finely honed. As we have seen, there is a propensity in peacetime to overestimate one's own striking power. Some extra firepower should be added when planning offensive combinations.

A tactical reserve must in fact be a safe reserve. Withholding part of one's missile capacity for a second strike presumes the survival of the ships with that capacity until they are needed. On the other hand, to be completely safe a reserve may have to be located so far to the rear that it cannot influence the battle, in which case it is not a *tactical* reserve at all.

A major consequence of massing for defense is the certainty that the enemy will be aware of the fleet and its general location. Then, electronic-warfare tactics should be designed not to mask the presence of the fleet, which is impossible, but to complicate the enemy's efforts to track and target the key units carrying out the fleet's mission—in a word, its striking power. In particular, no great sacrifice of fleet defensive firepower should be made to avoid detection and tracking. Active jamming and radiating decoys are the principal tools with which to inhibit targeting during this kind of overt operation. If the estimate is that the fleet cannot be safeguarded by its active defense long enough for it to attack effectively, then the whole operation should be reevaluated. Nothing is worse than a badly conceived plan that calls for the massing of defense and then destroys the effectiveness of the defense by an overly strict search and fire-control radiation policy. Powerful defensive warships such as Aegis cruisers are an electronic liability unless the force is powerful enough to attack overtly.

In the future, battles between moving forces will be fought at closer ranges than we expect because of scouting inadequacies and antiscouting effectiveness. Other battles will be fought with Arleigh Burke's metaphoric cutlasses, for missile magazines sometimes will have been emptied in panicky attacks.

Is there a way for a tactical commander and his staff to think through the exceedingly messy tactical problems of modern combat?

I offer the structure in the following section as a guide; it emphasizes scouting, C^2, and the two-sided nature of the battle.

A Force-on-Force Model of Modern Naval Combat

Let us first establish the purpose of the model: to help a tactician relate the scouting and weapon effectiveness of his force to that of the enemy so that the net deliverable striking power of the two sides may be compared. The model indicates the circumstances that govern which side will be able to attack effectively first.

Our previous models that compared firepower effects alone no longer suffice. Sensor-search effectiveness must be regarded as equally important. Emission control (EmCon) policies, which govern the extent of sensor radiation, are deeply involved. So are the distances between friendly and enemy forces.

These additions complicate the analysis but they are unavoidable, for effective scouting and sensor employment decisions are of paramount tactical importance. Still, the model is an endeavor to display only the most significant ingredients of modern naval combat, and in the simplest way possible.

Model Description

1. Two forces, Blue and Red, each have respective striking powers that are described, for all ranges in any direction, as good shots per strike, and a striking capacity that includes good shots in magazines. We will use the previous definition of good shots but refine it, saying that the number of good shots is range-dependent. Red and Blue forces may be massed, or divided and distributed in units as small as individual ships.
2. Each force has defensive power in its soft- and hard-kill defenses (interceptors, AAW missiles, chaff, etc.). Taken together, the defenses will be thought of simply as a filter that subtracts incoming weapons, leaving a net number of good shots that hit. The possibility of running out of defensive weapons will be ignored.
3. Neither side can deliver weapons or be ready to defend against the enemy's weapons without scouting information provided by reconnaissance, surveillance, ECM intercept, or other informa-

tion-gathering systems. The weight of striking power, a function of range, and the defensive power filter will both depend on the amount of scouting information.

4. Scouting information to Red and Blue may come from active search or from passive intercept of the enemy's signals. Passive information is generally received at a longer range than active search information and has different tactical content.
5. The *content* of scouting information will be treated in three categories:
   a) Detection: knowledge of enemy presence, enough to alert defenses but not to attack.
   b) Tracking: incomplete knowledge of the enemy's location and disposition, sufficient to launch an attack, but at a reduced level of offensive weapon delivery effectiveness.
   c) Targeting: knowledge of enemy force composition in such detail that individual units may be targeted and the best possible number of good shots can be delivered as efficiently as possible, *for the range in question.*
6. Scouting performance, in terms of bearing and range of detection, localization, and targeting, is a function of the electronic emission control of the active side. Such EmCon may be:

   EmCon A: restricted (minimum or no) search
   EmCon B: curtailed (some) search
   EmCon C: unrestricted (maximum) search

   For some active scouting systems, performance is given in sweep rate, that is, in area swept per unit time. For other systems, performance is given as a probability or frequency of detection per unit time in a searched region. For the composite search, the probability density throughout the searched region is dependent on time and governed by the search pattern. When either the tracking or the targeting of a detected force is also required to attack, either additional time or scouting effort may be needed. In all cases, the information (on the existence of an enemy force; on its location, course and speed; or on the details of its disposition) must be reported, so the relevant scouting time includes the time required to place the information before the tactical commander. Scouting effort is neither easy to conceptualize nor easy to manage, but whatever way it is portrayed, its effectiveness will boil down to the amount of area swept, the accuracy of the results, and the time it takes to report.*

7. Passive scouting performance, also given in terms of detection range, localization, and targeting, is a function of *enemy* EmCon policy choices. EmCon B resembles a tactical plan to deny the enemy good targeting information against primary targets (notably aircraft carriers or flagships) through passive scouting.
8. *Net* delivered firepower as a function of range (effective striking power minus defensive power) reduces the defender's offensive and defensive combat capability after the attack is delivered. In effect, the defender's prior fighting power (offensive and defensive firepower and staying power) and his active scouting capacity will be reduced according to the number of hits suffered in the attack.
9. Each Blue or Red unit that is mobile may move, carrying along its firepower potential.
10. On-board sensors move too. Other sensors may be in motion (e.g., satellites) or fixed (e.g., land-based radar). Each force's scouting capacity may be thought of as an ability to cover an amount of area. Coverage is the detection, tracking, or targeting of an enemy in successively smaller areas, which are, respectively, the region of interest, of influence, and of control.† The model stresses the simultaneity of scouting decisions on both sides and the tradeoff between radiating sensors, which give both sides information, and those that do not radiate, which keep information from both. The model is concerned in the most fundamental way with scouting resources and their deployment. The battle outcome rests on information collected and denied before the first weapons are fired.

*For an introduction—unfortunately not simple—to scouting methods, see Koopman.

†A more accurate model in some respects might define scouting capacity as coverage over a single amount of area, but with successively longer periods of time to detect, track, or target within it. The reader may judge for himself which approach is more appealing after studying the example that follows.

11. Once enough scouting information is thought to be in hand, an attack is ordered. Mounting and delivering it takes time, which is measured in hours. An enemy attack may arrive before the order is executed, rendering it null, or the enemy's attack may arrive too late, in which case both sides will suffer.
12. Surviving forces may reattack after accounting for:
    a) Damage from hits
    b) Aircraft lost in an attack
    c) Missiles expended

An Example of Modern Tactics

Envision a naval force (Blue) attempting to close and attack a land-based complex of scouting and firepower (Red). Red also has two missile submarines at sea whose mission it is to attack Blue forces before they are within effective range of Red's base complex. To concentrate attention on the scouting duel and simplify the discussion, we will assume Blue's force is strong defensively and so appropriately massed in one unit. Although some off-board strategic scouting has established the enemy order of battle and disposition ashore, Blue must use on-board sensors for battle scouting. Red's scouting resources will be introduced later.

In this example conventional weapons are used, and the situation is one in which an American carrier battle force conducts an attack with missiles and aircraft against an enemy with offensive aircraft and missiles.* The carrier force's mission is to attack a base as part of a campaign for sea control—that is, to suppress or eliminate a threat to American maritime activities. Red's mission is to destroy Blue so that it can continue air and submarine attacks on the shipping of Blue's allies. These forces and missions are consistent with one another in the context of conventional warfare. If nuclear weapons were involved, the forces, missions, and tactical plans would be very different.

Blue starts beyond weapon range, at 1,800 nautical miles from Red's airfield/missile-base complex. Red's two missile submarines

*It is too easy to make Soviet synonymous with Red. Considering the firepower of many nations today, the example has much more general application.

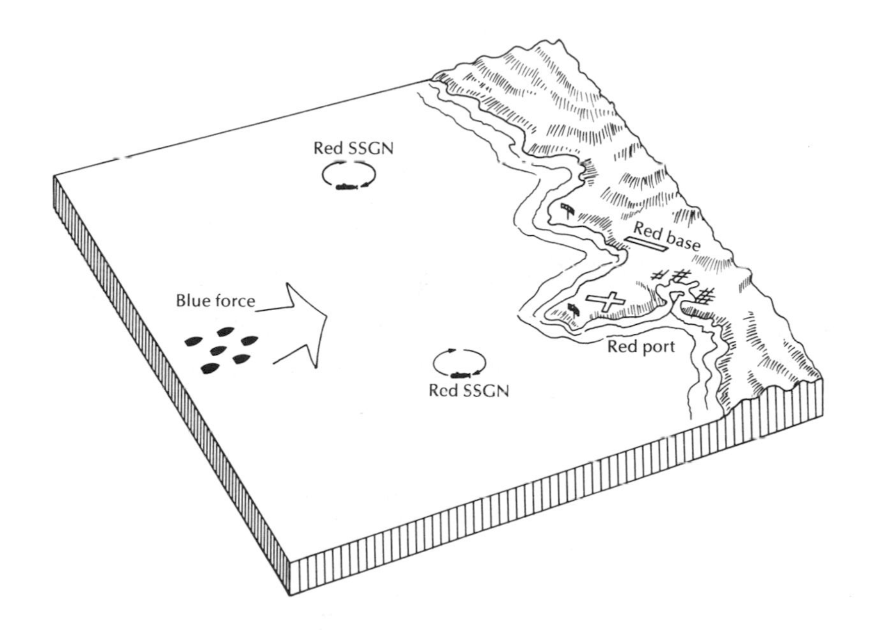

Fig. 10-2. Scenario of a Modern Naval Battle

are required by their defensive mission to stay at least 500 or 600 miles from Red's base, because Blue's striking power is very strong inside of 500 miles.

Blue's striking power (F) against land targets, in good shots (called shots hereafter) per strike, is represented in figure 10-3a. In this instance, the fifty shots that reach 1,000 miles are missiles. They may be fired once and are irreplaceable. The remainder of the Blue strike profile represents escorted attack aircraft that can strike repeatedly unless lost during the action. The full 150 shots by aircraft may be delivered out to 300 miles; beyond that, carrying capacity (and fighter escort strength) diminishes linearly to zero at 600 miles.

Red's land-based striking power of missile-armed aircraft is represented in figure 10-3b. Red can deliver 150 shots at short range, and his striking power decreases to zero at 1,500 miles. Red outranges Blue, but as we will see, he is not strong enough to attack effectively at his extreme range. Red may also strike again with any aircraft not lost in an attack.

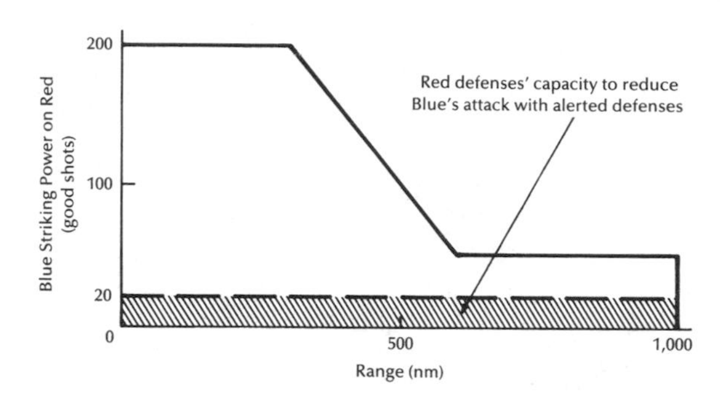

Fig. 10-3a. Expected Number of Shots Delivered by Blue on Red as a Function of Range of Red from Blue

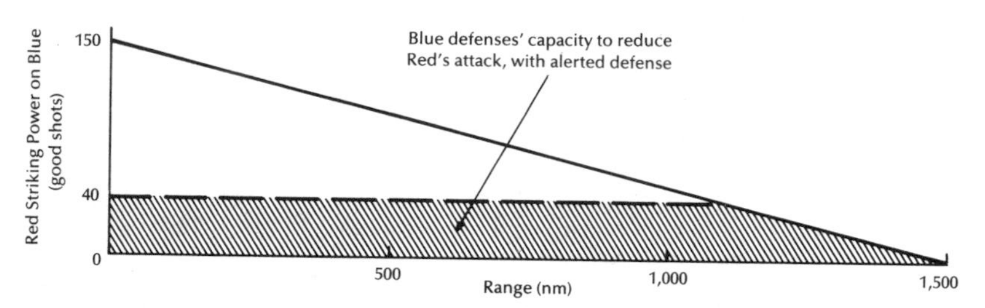

Fig. 10-3b. Expected Number of Shots Delivered by Red on Blue as a Function of Range of Blue from Red

Blue believes that when his defenses are alerted, his defensive firepower can eliminate forty of Red's shots. Therefore when he is outside of 1,100 miles he can defeat any Red attack, provided his defenses are given full tactical warning. Inside 1,100 miles some Red attackers will always penetrate Blue's defense and make hits. Blue estimates that Red's defenses, when fully alert, are able to take out the first twenty shots of his attack. Thus anytime Blue succeeds in launching his missiles inside of 1,000 miles, he is capable of doing *some* damage to Red. But—and this is key—a missile attack alone is not adequate, so Blue must close the range to, say, 500 miles and use his attack aircraft for a combined weight of attack, which at that range is one hundred shots, or eighty hits after Red's defense is accounted for. After that, Blue can reduce the submarine base and support facilities without significantly risking his fleet to Red's air base. For a conclusive attack, then, Blue may launch a coordinated missile and air attack from 500 miles, provided his estimates are correct and he has sustained no loss of striking power in the meantime, for he has no extra margin at that range.

What of Blue's ability to survive a Red attack? He estimates that his force can survive one hundred hits from a Red air and missile attack. Thus at the range at which he can deliver a decisive attack he himself is subject to a crippling attack: at 500 miles Red can deliver one hundred shots, of which Blue's defenses can handle at most only forty. His force will suffer sixty hits and be reduced to forty-percent effectiveness. If Blue is attacked first by Red with Red's full strength at about the critical range of 500 miles, his remaining striking power will be reduced to only forty shots, of which Red may be expected to defeat half. Twenty hits by Blue will not do Red enough harm to gain air supremacy for Blue, nor will it whittle down Red's next attack very much. Here is a summary of Blue's estimates:

Blue Forces	
Defensive power, fully alerted	40 hits averted
Staying power*	100 hits
Red Land-Based Forces	
Defensive power, fully alerted	20 hits averted
Staying power	80 hits

But a battle is dynamic competition, and we have not established Blue's maneuver potential. A reasonable sprint speed for his force is twenty-five knots. After he is attacked with Red's full strength at 500 miles, he may have six hours' grace before Red can mount another concentrated attack, in which case he may be able to close the range to 350 miles with his survivors and conduct a strike. At that short range his attack is sixty percent more powerful than at 500 miles; formally, his surviving strike capacity at that range is seventy shots.

*The fraction of scouting and firepower destroyed by enemy attack is the number of hits received divided by the staying power of the force.

After Red defenses take out twenty of those, the fifty hits achieved on Red will eliminate five-eighths—more than sixty percent—of Red's follow-up attack potential, giving Blue some modest hope of handling Red's now-reduced strike capacity. Not very promising? Consider the alternative, an attempt to withdraw. Blue flees as he is able, and Red reattacks at 650 miles at full strength. Blue faces an attack worth eighty-five shots. With only forty percent of his defenses remaining, Blue's defenses take out only sixteen shots, and he is defeated by sixty-nine hits against a residual staying power of only forty. Blue's situation is impossible if he tries to escape a reattack. Somewhere around 750 miles, Blue crossed the Rubicon.

What do Blue planners conclude? Nothing, until they have assessed the scouting capabilities of both sides. After all, much of Blue's tactical strength lies in his ability to move and in Red's lack of maneuverability. We will also see that Blue has a better way to combine maneuver and firepower—by winning the scouting duel—if he can offset Red's advantage in weapon reach.

Blue's EmCon plan must exploit the fact that Red's position, except for his two submarines, is fixed. Since Blue is a mobile threat, Red must search for him, and Blue's tactics involve the exploitation of Red's necessarily active search. If Red has enough satellite surveillance (or surveillance from any other scouting system that can covertly track and report Blue) to launch an effective large-scale land attack first, Blue will have to mass additional *defensive* firepower* (his offensive punch we know by now to be adequate) and radiate in EmCon B sufficiently for his full defenses to play. Blue's only alternative, barring a reckless hope that Red is less strong or more inept than intelligence estimates, is to abandon his plan to mass and try to attack in dispersed units—not a promising prospect, given the circumstances.

If, however, Blue has the ability to evade Red's long-range surveillance, he will be able to start his run in without having been located. In this case, let us assume that Red has two active scouting

*How much additional defensive firepower? Within the context of this paragraph, doubling his defensive firepower to destroy eighty hits per attack, perhaps less, would be adequate.

threats.* The first is an over-the-horizon radar, a surveillance system that blankets the ocean, giving a high probability of detection to a range of 800 miles. Blue must assume that the radar is active and that at an 800-mile range he will be detected, tracked at once, continually targeted, and subject to land-based or submarine attack within an hour or two. The second scouting system is a set of long-range "Grizzly" reconnaissance aircraft. The Grizzly reconnaissance effort can reach as far as 2,000 miles, but the longer the range, the narrower or thinner the search.

As Blue starts to close the range he will stay electronically silent, relying on his passive detection of the Grizzly's radar to give an attack warning in time so that he can shift to EmCon B, which allows his defenses to work at full effectiveness while still denying Red's scouting its full targeting potential. What will the probability of detection by the Grizzlies be? Both sides should perform extensive calculations, realizing that the range and breadth of the search are important variables under Red's control. Blue estimates that Red intends to attack inside of 1,000 miles (when a Red attack will achieve some hits) and beyond 600 (at which range Blue's threat to Red begins to mount precipitously). A Red attack at more than 1,000 miles may actually appeal to Blue, and one plan (which we will not pursue further) would be for Blue to stay at long range in hopes of inducing Red to attack where Blue has an advantage. Running in silently and swiftly, Blue will try to pick an approach bearing that reduces the chances of detection, and he may also use radiating deception units to divert the Grizzly search. Now, a reasonable estimate is that the Grizzlies will search out to a maximum range of 1,500 miles. It will take Blue twenty-four hours at twenty-five knots to close from 1,500 to 900 miles, and another sixteen hours to close to 600. Even in this simple example, the scouting situation is complicated enough to call for the diagram in figure 10-4.

By the time Blue is first able to attack—at 1,000 miles—he has a fifty-percent chance of being undetected. At 900 miles the chances

*A third, picket submarines, is not considered now, on the assumption that Red wishes to keep his two SSGNs closer in for missile attack. The possibility of submarine reconnaissance by Red can lead down many paths, one of which we will explore in due course.

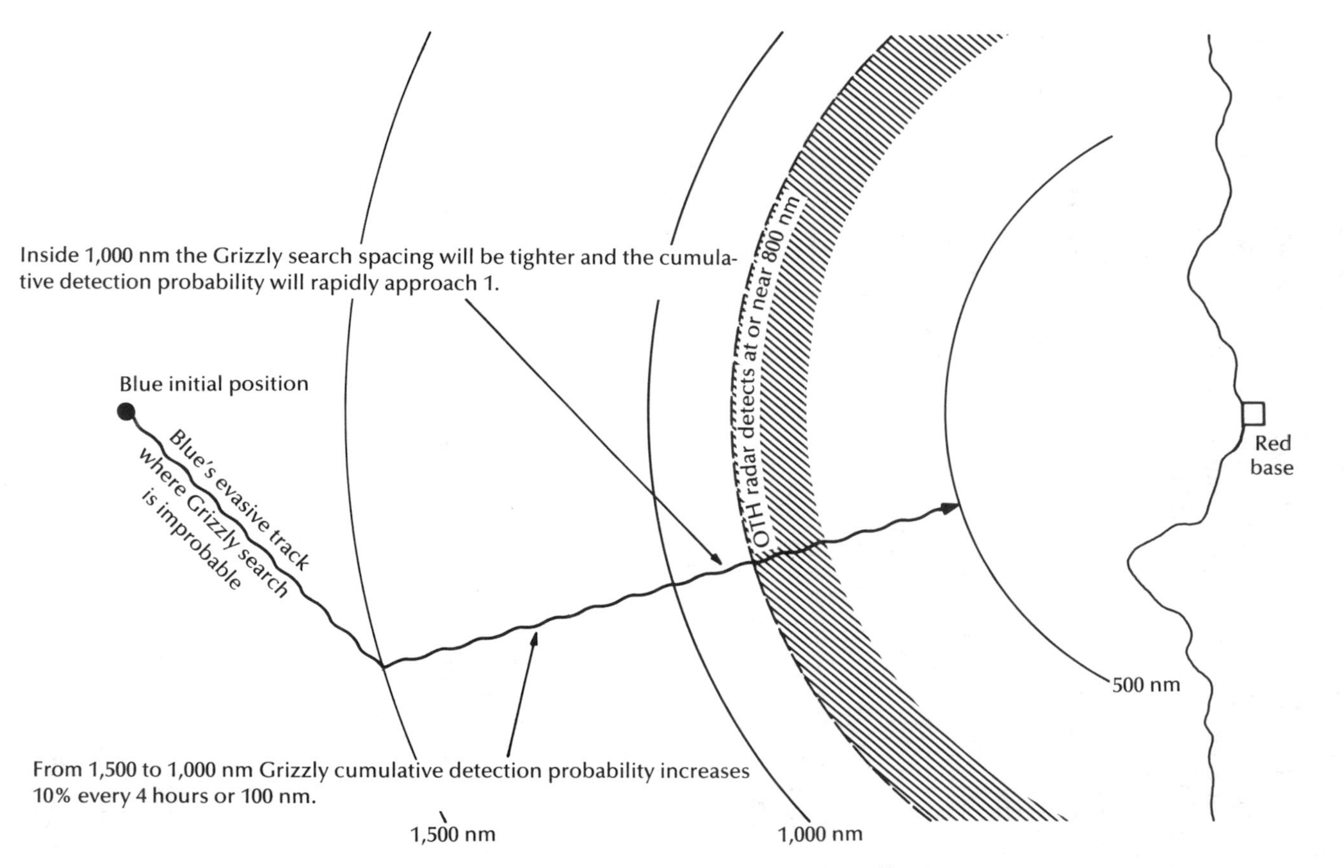

Fig. 10-4. Diagram of Red's Composite Scouting Effectiveness

of his remaining so are less than forty percent. At 800 miles detection is certain. Blue has, however, certain advantages. If he is detected by a Grizzly outside of 1,000 miles, he will know it and can prepare to defend effectively against Red's attack, or if he does not have faith in his defensive firepower, he can safely cancel or defer his attack. If he is detected inside of 1,000 miles, he will also know when, and will have about two to four hours to launch his own attack before the arrival of the Red attack. The resulting exchange of attacks will do heavy damage to both Red and Blue, the details of which it is not uninstructive for the interested reader to work out for himself.

There is another possibility. To see it, refer to figures 10-5a and 10-5b. Figure 10-5a depicts Blue's remaining striking power after a maximum Red attack from three distances, 500, 700, or 1,000 miles and farther. Obviously, Blue's residual striking power is much re-

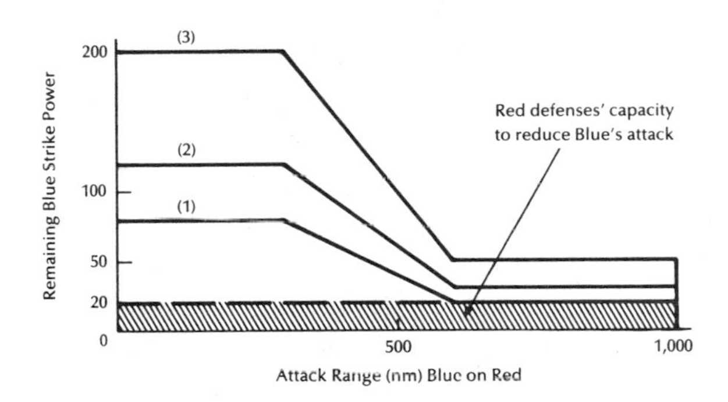

Fig. 10-5a. Remaining Blue Striking Power After a Red Attack at (1) 500 NM (2) 700 NM (3) 1,000 NM or more

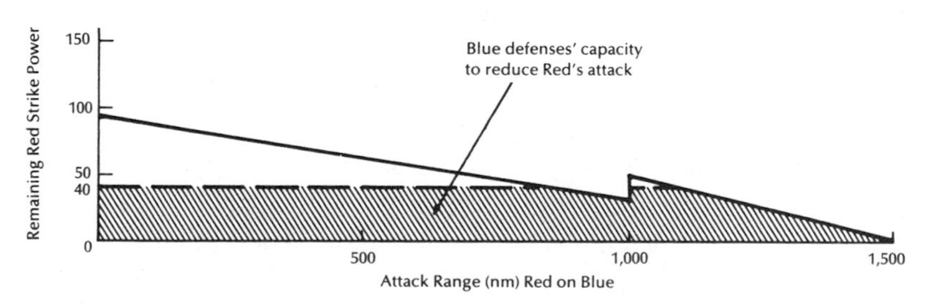

Fig. 10-5b. Remaining Red Striking Power After a Blue Attack from Anywhere Between 600 and 1,000 NM

duced after Red mounts an attack from the shorter ranges.The region above the crosshatching indicates the number of hits Blue could hope to achieve after Red defenses have defeated the first twenty shots. Figure 10-5b is a similar curve for Red. It shows Red's remaining striking power after a successful Blue attack, from anywhere between 600 and 1,000 miles. This figure indicates that although Red's hitting capacity is not destroyed by the Blue long-range attack, Red will be depleted enough that the Blue defenses can handle a Red counter-attack adequately to about a 400-mile range, as the next few paragraphs explain.

Thus, although a Blue attack from a range greater than 500 miles does not destroy Red's fighting power, an attack from anywhere inside of 1,000 miles is effective: it reduces Red's offense to near impotence. Therein lies the clue to Blue's best tactics for the combination of Red and Blue weapon and scouting capacities given in the example.

Blue should try to close to 1,000 miles undetected. As soon as he is within range, he should launch all his missiles. They will be an effective first attack. In fact, if he remains undetected during the two-hour transit by the missiles (it is assumed they are cruise missiles, with what would have to be, for conventional weapons, very precise terminal homing), Red will be taken by surprise. It is reasonable, then, to expect that nearly all of Blue's fifty missiles will be hits. But even if Red is alerted and the effective number of Blue hits is only thirty, three-eighths of Red's striking capacity is still knocked out. Had we not foreseen the effect of Blue's defensive firepower, the remaining five-eighths of Red's striking capacity would have appeared to be a serious threat to Blue.

Next Blue must close the range to about 400 or 450 miles. The reason is that, his missiles all being expended, he must get within his aircrafts' range. At that range he can launch a final, conclusive attack. During Blue's run in, Red will have counterattacked (he had twenty-four hours to do so). But Blue's defensive firepower is at full strength because of the tactical choice to mass his force, and because he will now be radiating all his systems overtly in EmCon C. Even if Red times his attack well and strikes at 500 miles, his reduced striking power will be only about sixty shots, of which Blue's defense will eliminate forty. Blue, with a survival capacity of one hundred Red hits, will be at eighty-percent strength for his air strike and capable of completely dominating Red as the end game to this example is played out.

So far little has been said about the two Red missile submarines. Submarines are usually thought of in their strike role. In this scenario Red places them at 500 to 600 miles, far enough out to attack and chip away at Blue before Blue's full striking power can be effective. The closer in they are positioned, the greater the possibility they will be able to attack. Now, we have never specified their striking power. If either one is able to target a carrier or Aegis AAW ship, its missiles might do enough damage to decisively undermine Blue's fighting power. Even a partially successful missile attack aimed against the force without complete targeting information might be enough to tip the balance in later engagements, when outcomes are poised on a knife's edge. However, Red's submarine tactics and their effect on the battle depend on submarine attacks made before Blue's main air attacks, a situation not unlike that in which Blue's best tactical plan is to launch his own early missile attack on Red's base.

Blue's need to make a twenty-five-knot approach enhances the Red SSGN's effectiveness. At high speeds ASW is difficult and surface-ship formations themselves are vulnerable to long-range (50–100 miles) sonar detections and submarine missile attacks. But a Blue decision to strike with missiles at 1,000 miles preempts Red's submarine tactics. The SSGNs cannot attack before Blue's missiles have struck, and after that they probably cannot cripple Blue enough to redress Red's disadvantage.

There is, however, another tactical employment for Red's two submarines, namely, as covert scouts. Blue's plan hinges on being able to approach in EmCon A (minimum radiation) and counterdetect the active radar search of Red Grizzlies in ample time to shift to EmCon B, permitting the use of fully effective defensive firepower. If Red puts his two submarines out to 1,100 or 1,200 miles as pickets, then Blue has to face a distinct possibility that he will be detected and tracked passively, and that a long-range *surprise* attack might descend on him before his defenses are up. The numbers are intolerable to Blue. At 1,100 miles a Red air strike can deliver forty shots.

Blue has counted on taking out the first forty shots with his defenses; it is for that very reason that he has massed his force. If he sustains forty hits because he is surprised, he has lost too much of his force to continue the battle.

The possibility of Blue's being tracked and targeted without his knowing it is a serious matter when Blue depends on Red's overt search to alert his defenses. If he must radiate for defensive reasons he cannot conceal his general location, and yet his tactical plan to mass is based on a strong defense. How serious is the threat of two submarine pickets detecting a battle fleet running in at twenty-five knots? At 1,200 miles, the probability of one or the other submarine detecting is perhaps one in four or five. In addition, the submarines must have the means to report a contact undetected. All in all, the odds of a Red covert scouting success are not high, but still Blue cannot escape the possibility of that success, which would lead to a devastating Red attack.

The biggest problem for Blue would arise if Red's major striking power were afloat and mobile. Blue's battle plan depends on knowing that Red can neither move toward him, which would rapidly make Red's threat intolerably stronger, nor away, which would render Blue's plan to launch missiles at maximum range ineffective. In the case of a maneuvering Red, Blue must establish his own active scouting plan, which changes everything. All elements now come into full play. Both fleets maneuver on a battlefield of grand dimensions. Two tactical foes struggle to devise scouting plans to find an enemy in motion and frustrate the enemy's plan. The fleets' striking powers are two coiled springs ready to snap into action when, for better or worse, either commander decides he has enough information and makes his fateful attempt to attack effectively first.

The Merits of the Example: A Summary

Now we must put the example in perspective. First, assume for a moment that the data were in fact not imaginary but as real and accurate as analytical methods can make them and as complete as circumstances permit, and that, in addition, all important variations or tactical options open to either side were explored. In other words, assume that what we have is the genuine, complete tactical analysis that feeds into a real battle plan. With all that given, the first enjoinder to every commander and his staff is nevertheless to stand back and ask where the uncertainties lie and what the margins for error are. For example, they should worry about Bernotti's grim observation concerning the effective range of torpedoes. What if Blue's missiles, fired at key targets on the ground at their maximum range of 1,000 miles, do not prepare the way for the air attack but merely alert the enemy? Until the first actual strike with missiles is tried, we are certain of nothing concerning their accuracy and effectiveness. A model of battle can beguile users into believing that it is more than any model can ever be. No one who knows naval operations is likely to make that mistake with the model I have used. But the danger is real that a complicated U.S. Navy decision aid, or any final, issued operation order, will be mistaken for a prediction. If the plan is sound it will work, but it may contain so many distortions as to be almost unrecognizable after the battle. Remember Nelson's simple-looking, tightly drawn battle plans for the Nile and Trafalgar and the wild, patternless appearance of the actual executions. Combat analysis does not aim to predict the future any more than a battle plan aims to represent the reality of the battle. Their objective is to help plan and win a victory. Analysis and plan are not sufficient, but they are necessary.

Our example does not portray real capabilities. It is intended to lay out, realistically, the special advantages warship mobility and maneuverability bestow on naval forces. These are precious assets when used shrewdly against an immobile enemy. It is obvious that if Red had been mobile, Blue's tactics, and especially his EmCon plans, would have disintegrated. If Red had had a force at sea larger than two submarines, so that Blue would have had to operate more sensors, the latter's whole plan of attack might have collapsed. The example also shows the real advantage of defensive massing. It does not show—though not much thought is needed to see—that against the scouting resources at Red's disposal Blue would have little to gain by dispersing his forces and much to lose. Successful scouting on Red's part would result in loss of concentrated attack for Blue and his vulnerability to sequential Red attack in detail.

A noteworthy shortcoming of the model is the way it distributes Blue's striking power and defensive force homogeneously among the unspecified number of ships in his fleet. Modern American firepower tends to be clumped together, striking power in a few aircraft carriers, defensive firepower in AAW missile cruisers and in the fighter aircraft aboard carriers. Such power is not reduced as gracefully as Blue's in the model. With two carriers in a battle force, a model result showing a fifty-percent residual aircraft striking capacity may conceal the fact that there is a twenty-five-percent probability that the force has no air striking capability (both carriers out of action) and a twenty-five-percent probability that the force still has most of it (both carriers operational).

The relative survivability of Red's land-based force and Blue's sea-based force, and the likelihood of their being reconstituted, is a difficult but vital assessment. Since the land/sea equation comes up repeatedly in American naval planning, even planning oriented toward sea control and protection of American maritime interests, survivability relationships must be handled with the expertise that comes from hard study.

By far the most important purpose of this example is to illustrate the processes—the dynamics—of modern naval combat. Even the most elemental analysis cannot avoid the messy scouting process. In the historical chapters we were comfortable describing the *essence* of naval combat with simple force-on-force attrition models. They helped demonstrate the structure of tactics and the correlations of force. They showed, especially, the decisiveness of a relatively small force advantage. By the time we came to World War II, however, long-range weapons were complicating our simple focus on attrition. Scouting had to be embedded in the attrition model. Offensive striking power looked more like a pulse than a continuing flow of destructive force. Staying power was more than ship survivability, and active defenses became important. So much for simplicity of tactical discussion.

This, in summary, is the way to think of modern battle:

— Two sides have offensive weapons, the potential of which is a function of range.
— Two sides have potential defenses.
— Each side has scouting systems, which must at least detect and often track and target the enemy for an effective attack.

— Each side's scouting activities may give away information that the other side will exploit to attack the enemy and defend himself.
— Each side has the potential to slow the enemy's scouting process by using cover and deception, inducing him not to use his sensors, or by taking other antiscouting steps, including attacks against off-board sensors.
— Finally, each side may interfere with the enemy's C^2 by either direct attack on the flagship or confounding his communications.

Ultimately the two C^2 processes govern all. Each commander's aim is to concentrate his firepower to achieve mission success. Concentration means the effective compression of the attack in time as well as its localization in space. Concentration is a focused pulse of destruction unleashed at the vital place.

As important as concentration is the timing of the attack. Throughout history, the genius of winning sea battles has not so much been knowing what to do as when to do it. This is still true. The crux of naval command is knowing when to commit the available attack potential.

Concerning weapon fire, modern naval battles will be fast, destructive, and decisive. Most often the outcome will be decided before the first shot is fired.

It is wrong for the tactician merely to maintain an offensive frame of mind, thinking of nothing more than getting in the first attack. Naval forces must execute the first effective attack—the one after which the enemy can neither recover nor counterpunch successfully.

Because modern battle fleets comprise relatively few units, the commander has the potential (not always the ability) to maintain tight control of his forces and, more so than the ground-force commander, to unleash the coordinated attack from widely dispersed positions. His tools for translating potential into ability are doctrine, training, a stable team, a compact system of signals, and a few commands signaled at the right time during the operation.

Each commander, faced with decisions based almost always on incomplete information, must decide how he can attack effectively before the enemy does. If one side, by all odds, is subject to *effective* enemy first attack, there is something wrong with its strategy, tactics, or weapons. You don't send a pikeman against an archer in an open field at noonday. But it is wrong to think of weapon range apart from scouting capacity; the two are wedded. Send your pikeman at midnight under a new moon. Any sound tactical plan must view the gathering of information on the enemy and the protection of one's own information as integral to the battle.

If the enemy's scouting and weapon ranges are both superior, then one's fighting strength, especially defensive firepower, must be vastly superior.

In cases when the issue is in doubt, a naval commander does not hold back a reserve. He tries to bring his entire striking power down on the enemy simultaneously. Otherwise economy of force is important. Even then he concentrates whatever he commits, and because of the fog of battle, he commits more than calculations show is sufficient.

A naval commander may have to commit weapons in advance of a concentrated attack to open up the way for it. He also needs some short-range weapons for follow-up. A battle whose outcome is decided is not over. The mop up—it may be conducted by either side—will be a scene of great confusion at shorter ranges.

11

Anchorage

Dropping Anchor

As a result of the wars in Korea and Vietnam I steamed around the world in destroyers two times. What struck me was not how much I saw, but how little. Our planet is a big place.

The visions of tactics in this book are likewise little glimpses of great vistas. We have circumnavigated the world of tactics, but as we come home to anchor we have neither seen nor charted most of its seas and shoals.

Of the insights this book affords, the following are the ones that stand out in my mind. They are offered as a few safe anchorages against the storms of battle.

Tactics

Naval battle is attrition centered. Victory by maneuver warfare may work on land but it does not at sea. At sea, first effective attack is the aim of every tactical commander. But grasping this is only a good beginning. Superior scouting, whose role has been underestimated in theory, practice, and history, opens the door for the attack. Command fuses scouting data, applies force, and frustrates the enemy's efforts to do the same.

The pace of battle cannot be captured in writing. Even war games have not communicated the sense of deliberate urgency that hovers over final events. Tactical success rests on timely action, well timed.

The goal of both tactical commander and weapon designer is to maximize *net* delivered offensive firepower. I emphasize net, first, as a reminder that staying power matters, but not to the detriment of the offense. As Vice Admiral S. O. Makaroff wrote, "A good gun causes victory, armor only postpones defeat."* No fleet can operate indefinitely in the face of a competitive enemy fleet without destroying it. I emphasize net, second, to draw attention to the fact that to be effective, weapons must be delivered—that ammunition in the launchers and magazines of sunk ships serves no purpose, and that weapons fired at decoys or a pinnacle of rock are lost forever.

Naval battles have been fought without a reserve. The object has been to bring the full weight of offensive firepower to bear in a compact pulse. Modern weapons may amend this single-minded tactical objective, but the principle of offensive concentration of force still holds for all naval components that are not specifically intended for fighting by stealth in solo operations. One old reason there has been no reserve is that the enemy's formation could be seen, which is most often no longer the case. The principal reasons now are the decisiveness of the first effective blow and the cumulative benefit of an initial advantage in delivered firepower, phenomena that still shape the face of battle.

Limitations on the ability of warships to operate for long periods within range of land-based weapons of comparable striking power are as severe now as they have always been. Contrarily, the advantages warships have over ground forces, namely greater strategic mobility and tactical maneuver, are probably as potent as ever. The tactical advantage of maneuverability that naval forces have over ground forces must be exploited: operating ships from the same position invites attack.

*Makaroff, p. 181.

Greater weapon range and lethality have expanded the modern battlefield. Of the many immediate consequences, one is that more land-based forces must now be considered maritime forces, and another is the need, manifestly more important than before, for sea-based forces to team with land-based forces. Successful cooperation in joint operations, both strategic and tactical, is a modern imperative.

The major fleet ambiguity that must be resolved, in present doctrine and in future technology, concerns the following:

— Reliance on active defense: massing enough force to defeat any probable first attack by the enemy. The fleet gives up moving "quietly" in order to fight defensive weapon systems cooperatively.

— Reliance on stealth: distributing force to make the enemy work so hard and take so long that he cannot attack effectively first. The force moves quietly, or the advantage of stealth is lost.

There will never be a final, certain preference for either active defense or stealth, because of swings between the power of the offense and the defense. Either choice gains and relinquishes advantage.

Fleet exercises designed for the opening salvo of a war under "rules of engagement" are graduation exercises—the most difficult to execute. It is a mistake to overemphasize the first strike of the war in battle practice. The Tactical doctrine is designed to guide the whole of wartime operations, and tactics for war's outset are a special, difficult case. Perhaps more ought to be said about the importance of winning the *last* battle of the war.

Firepower, scouting, C^2, and their antitheses all play their roles, but success comes from the synthesis of all six in the face of an enemy who is trying to do the same. A good tactical plan blends its parts like a musical composition, with subtle harmony, heroic solos, quiet intermezzos, and crescendos rising like thunder out of the heavens.

Weapons and Sensors

The evolution of firepower warrants careful study and deep understanding. When Lanchester's square law of simultaneous attrition best describes the battle, a 4:3 advantage in *fighting power* (the conceptual or mathematical synthesis of offensive and defensive force) will be decisive and a 3:2 advantage overwhelming. When pulsed power describes the process, a weaker side outnumbered in fighting power by 2:3 may win, if through scouting and C^2 it can attack effectively first. Since the beginning of this century, which saw the development of the torpedo, the latent threat has existed that an inferior force could defeat a superior one because of the ability of modern arms to deliver a highly effective pulse of firepower. In the modern missile age, this latent threat has achieved new status.

Tactical planners who expected long-range weapon fire to be effective have been astonished to find that, for any of a number of reasons, battles are still decided at much shorter ranges. In the past, the reasons had much to do with the pace of scouting and its assimilation while the two forces were changing their relative positions, or with the position of the land in a deadly game of hide-and-seek on and under the surface of the narrow seas. Commanders of future battles will need to be proficient in anticipating both sides' actual weapon delivery range, which is not necessarily effective weapon range and is rarely maximum range.

If an inferior force must fight for reasons of strategy, then:

— A scouting edge is mandatory. An estimate of the situation (a game theory solution) is made, but for the purpose of ascertaining how best to take risks.

— The solution should offer hope that with the edge from scouting and deception, attacking first will in fact win, i.e., be effective.

Warships can be constructed to survive and fight after taking several shell, bomb, and cruise missile hits, and more than one torpedo. More ships in modern fleets, which are built primarily for conventional war, should be designed for survivability and continued fighting. But warships designed for nuclear war fighting will not survive through built-in staying power.

Because a fleet action is so decisive, a great deal of attention will be devoted to maintaining superiority or eroding it. Today, as a consequence of U.S. maritime superiority, mining and mine countermeasures, which are peripheral to the study of fleet tactics and have tactics of their own, are likely to dominate wartime actions on both sides in surprising ways. Guerrilla warfare in the form of submarine attacks, launched either to erode the enemy fleet's superiority or to attack the enemy's commerce, can be predicted by anyone who knows history, and attacks for the same purpose by long-range aircraft can be predicted by those familiar with the present state of scouting and long-range aircraft technology.

Planning and Execution

Doctrine unites action. It influences and is influenced by training, technology, tactics, and objectives. Doctrine, the instituted set of procedures for combat, should be compiled for the people controlling weapons systems, ships and aircraft, elements of the fleet, and the fleet as a whole. These procedures must be compatible. Doctrine at all levels should be specific, designed to achieve the best results from a united team, but should also allow room for inspired tactics and initiative.

Doctrine standardizes tactics to reduce the laborious planning of individual operations. It is, in effect, generic planning for what can be practiced and trained without knowledge of specific mission context. Modern fighting instructions are needed, without the strictures imposed by the old Royal Navy's permanent fighting instructions. A battle plan is mission-specific doctrine.

The more unstable the force composition, the greater the need for standardized tactics and commands. Likewise, the greater the rate of turnover of tactical commanders and COs, the greater the need for personal consultation in planning and for tactical simplicity, although these cannot substitute for a unit's stability and cohesion. A modern signal book standardizes tactical commands. In action, tactical communications should be brief, unambiguous, and few.

In future war, exploitation of enemy signals is going to be very important.

Planning and execution are related, but they are not one and the same. Treating them as identical would be disastrous. One does not plan during execution; one executes a plan that incorporates tactical variations and permits alteration in the execution. The plan must be almost as firmly implanted in the captains' minds as the operation's objective itself. The plan cannot depart greatly from doctrine and training, and execution will falter without doctrine and training. Good execution may look so different from the plan that to the untrained eye there is no similarity.

Battle plans need to provide for contingency operations after damage, for example to the carrier, the best AAW ship, the flagship, or a vital scouting component (on or off board).

The single best piece of practical advice for a tactical commander may be simply this: Decide in advance what must be done and embed it in your commanders.

Relationships with Strategy

The bare-bones conclusion that naval combat is centered on attrition, which implies that death and destruction are unavoidable, is exceedingly gloomy. After that central fact of naval warfare is thoroughly grasped, four ameliorating tendencies may be considered. First, a show of superior force is more likely to deter a naval battle than it is a ground battle. Neutralization of an inferior force is a common historical phenomenon. It is obscured in the study of tactics because history records battles that occurred rather than those that did not. Second is the trend in combat toward a higher rate of destruction of machines accompanied by lower relative losses of fighting men. Third is the trend toward a greater proportion of noncombatant personnel performing scouting, antiscouting, and logistical roles at a remote distance from the battlefield. Last is that, since the consequences of declining decisive battle are less immediate and obvious at sea than on land, the nation with an inferior navy is less likely to fight a battle. An encircling navy is less threatening to the jugular than an invading army. This has been true when the weaker navy belongs to a continental power, such as the Soviet Union. If a maritime nation such as the United States is threatened by a superior navy, it is destabilizing, leading to arms races in peacetime and bloody sea battles in wartime.

Mahan was right to emphasize that it is the foremost responsibility of every battle fleet commander to concentrate forces and win battles with tactical skill. Corbett was right to emphasize that strategic considerations determine whether a decisive battle should occur. A fleet may serve its strategic purpose without engaging in a decisive battle.

Many things—submarine warfare, or the responsibility of safeguarding national interests worldwide, for example—may distract a battle fleet from its role of fighting or forestalling the decisive battle.

Fleet actions occur when some serious situation on land will eventually be intolerable if the weaker fleet does not come out to fight. The tactical consequence is that one commander will be burdened with unwanted baggage. While dealing with the enemy he will simultaneously have to protect shipping, troop ships, or a beachhead.

Every commander's staff ought to be able to tell him by the orders of battle who is superior and by how much, even in these days of asymmetric force compositions. Raw orders of battle, however, do not reveal superiority. What does so is a comparison of force, not of forces. Weapon weights and ranges, scouting capacity and C^2, including the tactics to which both sides are trained, are all variables in the equation. Modern decision-support systems are necessary for preliminary force comparisons. But these comparisons are mere abstractions. The actual deployment decisions are what incorporate missions and geography, and what determine force assignments for alternative tactical plans. The final comparisons of deliverable firepower are adjusted accordingly.

The correlation of forces is a good predictor of victory. This statement must not be taken for more than it is. A prediction may fail because of the unforeseen exigencies of battle. Predicted losses are likely to be greatly amiss. The statement also must not be taken for less than it is. It is folly to dismiss the need for computations merely because of their uncertainty. No military planning can proceed without calculation and prognostication.

Clausewitz wrote that the concern of an army without a mission is self-preservation. Just so for a navy. SSBNs should hide in conventional war. Carrier battle groups should be out of the way at the onset of nuclear war. Both may have vital roles to play later. Self-preservation has been the observable aim of navies too weak to enter decisive battle at once.

It is a historical fact that influential navies have always been expensive to build and maintain. But the worst of economic policies is to build a substantial navy that in the end cannot compete against the enemy.

Some uninformed commentators say there can never be another old-fashioned fleet action because of modern weapons and sensors. Perhaps they have obsolete tactics and ships of earlier wars in mind. As long as nations use the surface of the ocean, they will struggle to maintain fleets that can control the surface, and those fleets will include surface warships.

The Study of Tactics

One thesis guiding this book is that tactical military study strives to bring whatever order and understanding is possible out of the chaos of battle. I have supposed that the foremost intellectual way of doing so is to estimate the naval effectiveness of chosen tactics for battle, and to offer the resulting measurements as a practical guide to the tactical and strategic planner. Some usual measures fail, namely, (1) performance measures, such as detection range, (2) operational offensive firepower, such as the number of offensive missiles in the launchers, (3) comparisons, such as Red's offensive firepower versus Blue's defensive firepower, and (4) values for ASW, AAW, or strike-warfare duels treated in isolation. The measure that will be useful for planning is an estimate of the losses both sides sustain while accomplishing their missions, taking into account specific scouting plans and firepower delivery intentions. There are two cautions, both seemingly paradoxical:

— While the results of the measurements are predictors, because of the residual chaos of battle, the results are not predictions.

— While the measure of effectiveness I recommend is the losses to both sides, this may be a proxy and only remotely linked to the tactical commander's true mission, which sometimes will have nothing directly to do with attrition.

It was the best tactical thinking that paid off in World Wars I and II. And that thinking was done by military men. It is not possible for sound tactics to be developed by anyone else. Inspired new ideas can come from elsewhere, yes, but the acceptance and working out of tactical details must occur in the fleet. The fleet, however, is preoccupied with administration, weapon readiness, and engineering reliability. Tactical improvement requires the assistance of fleet schools. And the best tactics will only result when the teachers are the best tactical officers, and when it is understood by everyone that their role is not only to teach but to improve tactics.

Historical analysis of naval battles will help men explore and calibrate cause-and-effect relationships, which are the foundation of modern battle models. Studies must analyze positions, communications, and tempo in detail, and be as rich in data as in narrative. Studies should emphasize the six processes of combat and their interrelations in a force-on-force paradigm. The tactical scouting process, as it was carried out from World War I onward, should receive careful attention.

History demonstrates that naval weapons perform phenomenonally below their peacetime potential and expectations. Why then the paradox of decisive battle?

— A tactical reason: ineffectiveness can be two-sided. When gunnery failed Sampson at six thousand yards, he was able, because of his enemy's ineffectiveness, to close to one thousand yards.

— A technological reason: weapons are highly lethal. After many missiles miss, one hit will be enough to render many a ship *hors de combat.*

— A strategic reason: history devotes itself largely to battles that have great consequences. The study of indecisive battles is just as instructive tactically.

Some focus on the science of war, others on its art. I would rather approach the study of war from a different perspective, highlighting "command mystique," that is, the quality of spirit that distinguishes brave, wise, and inspiring leaders. If this third perspective is given its due, I doubt that any remaining differences between tactical science and art will seem very important. What will abide is a sharper appreciation of the fact that good practice grows out of good theory, and that both are necessary but insufficient for consistent success in battle.

Epilogue: The Next Battle of the Nile

We are in the *Orel*, in the flag operations center of Vice Admiral Pyotr Ossipovitch Briam. After the abortive Southeast Asia campaign of 1996, our leader saw the folly of commanding ashore from TVD headquarters at Cam Ranh Bay. For five weeks he has been guarding the eastern Mediterranean north of Alexandria. Three levels above, the summer sun beats down, its heat mollified by a brisk twenty-two knots of wind across the deck, product of a gentle breeze off the bow and fifteen knots of speed generated by our humming nuclear propulsion plant. Here the machinery sounds are muted and all is quiet, tightly woven efficiency.

To the consternation of NATO, Briam has safely delivered Comrade General N. V. Bonapov to Egypt, and that magnificent leader, the darling of the Central Committee, is now deep in the Sudan, sweeping aside all opposition. It is the last step in a masterful maritime operation that has isolated the Near East and is destroying the moral fiber of the West at a single blow, completely reversing the effects of the Israeli-Syrian-PLO accommodation of 1995, which threw the entire Middle East into the arms of Western capitalism and sent shock waves of despair through us all.

Now, on 1 August 1998, we can fully appreciate the brilliance of Comrade Bonapov's plan. We still remember our stocky general, his eyes flashing, striding his cabin as he described how he won over the Presidium to his daring plan. Greece and Turkey were frozen by their own antagonism after Greece withdrew from NATO. "After I foment a crisis in the Caribbean while there is turmoil in West Africa," he said, "we will rip the Sixth Fleet right out of the Mediterranean. Into that power vacuum Briam will carry my mobile army, pushing through the Dardanelles before the Turks can think about it, and killing the Montreux Convention once and for all. Rattle the nuclear saber in its scabbard and the whole West will be paralyzed and in disarray. The key is the Sixth Fleet. Clear it from the scene for just thirty days, *Migom,* and I will have Egypt and its airfields. All of southwest Asia will tremble, unsuccored by the NATO paper tiger, threatened by you from the north and by me from just across the Red Sea in eastern Africa."

And so it has come to pass, a force of thirty-six thousand landing at Aboukir Bay on the first of July and sweeping south. Turkey is mobilized but isolated and impotent. Greece, the spoiler, is dangling by its own countrariness. Italy is wracked with uncertainty, its NATO membership hanging by a thread. Israel and Syria are howling in frustration and fear. In Amsterdam the people riot to take the Netherlands out of NATO before the United States—not us!—fires a crazed nuclear strike. In New York City, street fighting rages in front of the UN for the benefit of live television, half of them yelling for suicidal total war and the others for suicidal total peace. We don't need spies to tell us that Washington is behaving well, and just as Bonapov predicted: there is a strategic alert, of course, but coupled with hysterical reassurances that the nuclear sword is sheathed, and the Sixth Fleet, the West's only hope, has been dispatched hastily back across the Atlantic for the Strait of Gibraltar.

Thus Briam's fleet is the lynchpin of Bonapov's plan. Our admiral has been successful in every way. The stunned American submarines have been driven out of the eastern Mediterranean. Their flimsy solo tactics have been no match for our two-pronged ASW defense. Perhaps they are held at bay by our active bottom-bounce sonars, which triple the risk of an approach and torpedo attack. Perhaps it is our *Tovarishch* tactics, our deadly hunter-killer teams. Best of all, our carriers with their STOVL aircraft have been adequate enough to provide air cover and striking power, supplemented increasingly now by tactical aircraft out of Egyptian airfields.

We wish, of course, that Admiral Briam had more experience at sea. Handpicked by Bonapov, he is, all concede, a master strategist, standing at the right hand of Comrade Bonapov in all his careful planning. And he is master of the supporting systems that calculated the correlations of forces, gamed dozens of variations, and computed the timing of every move. You say our admiral is shy of naval combat? Who, comrade, were the alternates? We are a great land power. Afghanistan, Cuba, Nicaragua, Czechoslovakia, Hungary—these were incidents that gave our navy men no fighting. Briam is the best we have: solid Great Russian stock, bold and deep.

Still, he is notably nervous about the prospect of the return of Vice Admiral Grant and his Sixth Fleet. This descendent of a former American general is a reckless madman, heedlessly thirsting for battle, and his idiot seamen follow him like children. Briam was too hasty in the trigger-happy expenditure of forty-seven SAMs against the four surface-to-surface missiles that came at us from the Lebanese coast; a few such deceptive attacks and our magazines will be bare. And did he abandon wisdom by moving out from the coastline? A week ago he took our big ships westward, away from Lebanon and Turkey. Now neutral Crete is on our northern flank, friendly Libya on the southern flank, and so on three sides the fleet is free from land attacks by missiles or aircraft. The operations officer cautioned that we should not have sent so many of our AAW ships so far to the west. It is not *doctrine,* he protests. But Briam wants them well out in front to surprise the aircraft from Grant's onrushing Sixth Fleet. He has plucked out all of the enemy's eyes and filled his ears with wax. To him, our footsteps are like a ghost's, our voices less than whispers.

Yes, Comrade, and the glory is that our own admiral can see through blackest night. He says to fight the American man of surprises takes surprises of our own, and our comrade leader has waiting for this Admiral Ulysses S. Grant the surprise of a lifetime. For even as the Sixth Fleet steamed west toward Cuba the last twenty-fifth of June, our destroyer *Biedovy* stopped the Greek trawler *Nemesis* for inspection while the blacked-out little ship was suspiciously loitering off the Hellespont, between Gallipoli Peninsula and the ruins of Troy, auspicious scenes of two prior debacles in Western civilization. Catch of catches, the Greek ship was a lie, a deceit, full of American spies and listening gear and carrying Grant's private crypto equipment of the most ingenious kind. It operated on a Hewlett-Packard computer linked by satellite directly to his flagship. Briam feeds Grant just what he wants him to hear now. Whatever his operational shortcomings, Briam is a master of *dezinformatsia.*

Grant will fall to signals warfare and counterintelligence, the Americans' weakness—they are an open book. We have the file on this man, who did us so much mischief at the Battle of Tanjung Pinang in 1997, when he was merely a commodore. That was the occasion in whose aftermath the iron disciplinarian Rear Admiral D. D. "The Dirk" Porter said, "How can you accuse Grant of disobeying my orders? He did exactly what I would have told him to do if my communications hadn't been cut off by the enemy." Ah, yes, we should have had them both and saved Cam Ranh Bay. Young Grant is the very center of gossip in Georgetown for his notorious attachment to the Princess Dewi of Bali, whom he met after saving Indonesia. In her garden, it is whispered by everyone, they were Ulysses and Lady Helen, and he read to her the poetry of Homer and Virgil.

Here is another intercepted signal sent by Grant. It confirms our intelligence. Three days ago he made his penetration of the Strait of Gibraltar and our first trap was sprung. One carrier was damaged and is limping to Barcelona. One Aegis cruiser was sunk. We know these details from his very own circuits. The battle cost him heavy expenditures of missiles, too. *Aga!* But we also have our magazine problems. These nagging surface missile attacks from the shore have depleted our magazines and replenishments are hard to push through.

Turkey? No, it is not the Turks at the Dardanelles that interfere with our logistics. Here is where this Grant has outwitted us—I admit it. It was after the fishing boats and coastal shipping started creeping back into the eastern Mediterranean, cluttering our radars and confounding our surveillance picture. We should have shot them up, you say? No, this time our comrade admiral was right. Men will fight for food and fuel, and the Greeks and Turks and Lebanese must have full bellies and warm homes and cars with petrol in the tanks. Anyway, we do not have enough weapons to sink them all or enough surveillance to track them all. That is what Grant saw when he slid those absurd little eight-hundred-ton missile boats into the Aegean islands. Even the Americans call them Wilson's Follies, after the inept secretary of defense who said, "I will show the United States Navy how it is going to get its six-hundred-ship navy." *Cushings*, he called them. Who do you suppose this Cushing was?

So now Grant has them sprinkled like nettles throughout Greek waters, ducking in and out of the islands, camouflaged by day and popping their old-fashioned missiles at any badly armed or unprotected ship flying the Red Star. Who would have thought we would lick the American submarines and still be escorting convoys through the Aegean because of these spit kits?

Ne bezspokoytes. The impatient Grant will push his battle fleet into the eastern Mediterranean in the next few days and Briam will have him. The Strait of Sicily is a deathtrap of mines, submarines, and air-, surface-, and land-launched missiles—like nothing the Sixth Fleet has ever known. The Pillars of Hercules were child's play compared with the Scylla and Charybdis we have prepared for this American Ulysses.

Briam has a certain spring to his step. In his hand are six messages, all decrypted from the Sixth Fleet's command net. The American formation is plotted on the telescreen, the last signaled position and intended movement—the American PIM—plotted in Briam's own combat direction center. Surveillance radar from Libya, fortunately having survived the desultory air attacks of the unguided, old-fashioned American bombs, confirms the formation and position.

At Briam's elation our spirits soar. "In one hour the reckless Ulysses enters the straits. In six hours he is in the minefields. In eight hours I will strike from every direction. He can hit me only from the west; I will hit him from east and south and west. Now we attack first! Send: Execute Plan *Pyotr Velikii. Otomstim za Bitvu Pinang. Unichtozhit!*"

It is this very moment that the first of three fateful messages arrive. It flashes on the screen. Briam frowns and turns. "Before it was destroyed by precision guided missiles from American A-18s," he reads gravely, "the Libyan radar reported evidence that the Sixth Fleet had reversed course. What is this Grant up to?"

The second message is delivered by the hands of a puzzled chief of staff. "I have from Moscow that Grant is not with his fleet. This report says that for two days he has been somewhere in the Aegean."

For a moment Briam stares in stunned silence. Slowly the blood drains from his great ruddy face. We follow his shrewd brown eyes as they search carefully across the displays. "We do not wish Grant to be to the north," he says. "My defensive posture is oriented to the west. But what is the threat? There is no threat."

"There are the *Cushing*s." It is the operations officer speaking.

"They have no eyes," says Briam. "We have put out all their eyes. *Cushing*s have surface search radar only, range twenty miles. They are far to the north. They would be shooting blind." He pushes the automatic relay of the surface search radar. Staring out at him is the coastline of Crete, sixty miles away, as though drawn on a map. "*Bozhe moy*. There is a summer duct. We can see on the surface for a hundred miles. And so can he."

Even as Briam speaks, the special intelligence officer shoves a path through us, saluting with his right hand while the left thrusts a hastily scribbled message forward. "Comrade Admiral, I have an impossible transmission, intercepted on the Sixth Fleet private channel. I cannot translate it!" This from our intelligence officer who puts himself to sleep with Shakespeare.

Briam reads it. "*Sukin Syn,*" he utters. At his oath, the hot flush of palpable fear radiates through the flag center, spreading paralysis everywhere. It is in all the faces. "*Somebody* translate this abomination," he thunders. The message flutters before our eyes.

> For Admiral Pyotr Ossipovitch Briam:
> *Equo ne credite, Teucri. Quidquid id est, timeo Danaos et dona ferentis.*
> Your Servant,
> Ulysses

Briam's eyes stare at the surface radar, and we tremble behind him. "He knew that we knew. Every word of traffic on Grant's special channel has been *pustayaka,* the voices of the Sirens." Our titan is about to fall from the heavens and we with him. On the old-fashioned surface radar screen little pips move out from behind Crete, sixty miles away.

Tiny specks dart rapidly from the pips and disappear above the surface duct, too rapidly to count, but we know there will be eleven or twelve from each pip because a *Cushing* class carries twelve Harpoon IIs. These surface-to-surface missiles are almost obsolete, but they will home on any ship afloat at which they have been fired. A solid, overwhelming sheet of one hundred or more of them will strike us simultaneously and almost unannounced in about ten minutes. . . An Olympian thunderbolt from out of the undefended north!

"What is the effective range of their weapons?" asks Briam.

"Eighty miles, Comrade Admiral."

"*Da,* I thought as much. *Konechna*. . . Of course."

"Sound battle stations!" cries the flag captain.

Shaken from his paralysis, the chief of staff screams, "Warn the fleet!"

With the fatalistic calm of a Great Russian, Briam says, as if to himself, "It will be too late."

Admiral Grant, the new Ulysses, master of timing and delivery of fatal blows, strikes his last on Admiral Briam seconds before the missiles arrive. A wizened man somehow left behind from Bonapov's entourage of Africa-bound Russian scientists knows Latin. He translates the message, which turns out to be from the Aeneid, and delivers it in a quavering hand:

> *Trust not the horse, men of Troy.*
> *Be it what it may, I fear the Greeks,*
> *Though their hands proffer gifts.*

Appendix A

Terminology

The use of terms in this book is consistent with the definitions given below. Wherever possible, standard definitions from the *DoD Dictionary of Military and Associated Terms, 1979* (JCS publication no. 1) or from standard dictionaries have been adopted or adapted. Soviet definitions were also consulted; tightly drawn though they are, they could not be adopted *in toto*.

Military terminology varies among sources and is casual in application. Even where agreement is widespread, I have found it impossible to apply "standard" definitions everywhere. Most dictionaries, for example, define tactics as an art and science. This is not general naval usage. In nautical language tactics are procedures, actions, things done or directed to be done. The science of tactics is just that—the science of tactics.

Doctrine is one of the military's most elusive words. The U.S. Navy has usually avoided the problem of defining the term by ignoring it. This is unfortunate. Doctrine as a concept and as a practice should be carefully delineated and put to work.

One of the most common and potentially disciplining sources of military terminology is physics. Even though such concepts as power, energy, pressure, and momentum cannot be as quantitative or unambiguous in the social study of men at war as in the study of inanimate objects, usage need not be as careless as it is in military affairs. For the purposes of this book, I have had to define only *force, power*, and some of their compounds. For physics terms applied in a military context, the definitions offered below are, I believe, the best.

Special note must be made of the word *counterforce*. In the lexicon of nuclear planning, it refers to offensive attacks on enemy forces. As a prefix, *counter-* indicates opposition in direction or purposes; as an adjective, opposing or contrary; as a noun, an opposite or contrary thing; and as a verb, to oppose, offset, or nullify, to meet attacks with defensive or retaliatory steps, or to act in opposition. Thus usage with a defensive denotation seems best, despite the fact that defense against nuclear weapons has only been possible by offensive attack. I have chosen to restrict the meaning of tactical counterforce to defensive measures—that is, measures used to defeat an enemy attack in progress.

General Military Definitions

BATTLE. A general encounter, which includes combat, between opposing armies, fleets, or many aircraft to achieve conflicting aims.

COMBAT. Conflict involving the delivery of lethal force between opposing sides in a fight, action, engagement, or battle.

CONFLICT. Competitive or opposing action of incompatible forces.

DOCTRINE. Policies and procedures followed by forces to assist in collective action, either strategic or tactical. In a broad but acceptable sense, doctrine includes battle plans and practices for the immediate application of force.

STRATEGY. Policies and plans that govern actions in a war or a major theater of war. (Strategy establishes unified aims of war and sites for the employment of forces allocated toward those aims. The intention of strategy is to affect the outcomes of wars or campaigns, of tactics, the outcomes of battles or engagements. Therein lies the distinction and connection between them.)

TACTICS. The handling of forces in combat; acts of deployment, maneuver, and application of force. (Sound tactics are procedures that employ forces to attain their full combat potential. It is not possible to define tactics or sound tactics as procedures to *win* a battle.)

The Elements of Combat and Related Definitions

ANTISCOUTING. Actions taken to destroy, diminish, or preclude enemy scouting effectiveness. (Antiscouting includes the destruction of enemy scouts, such as shooting down a surveillance satellite or reconnaissance aircraft, deceiving enemy sensors, jamming sensors to reduce tracking or targeting effectiveness, and interfering with a scouting report.)

COMMAND AND CONTROL (C^2). Decisions made and actions directed by the commander to employ force, counterforce, scouting, and antiscouting resources to accomplish an objective. (C^2 includes the integration of scouting information, combat decisions, and the dissemination of these decisions, but it excludes acts of scouting themselves. Support for C^2 includes staff work, decision aids, and communications systems.)

COMMAND AND CONTROL COUNTERMEASURES (C^2CM). Actions taken to defeat or delay the effectiveness of the enemy's C^2. (C^2CM includes destruction, communications jamming, and the intrusion of false communications. Signals exploitation, however, is most appropriately categorized as an act of scouting.)

COUNTERFORCE. The capacity to reduce the effect of enemy firepower. In this book, it is the aggregate of defensive force and staying power. (Although not so used here, counterforce can also include offensive attacks against enemy forces.)

COVER. Secrecy, camouflage, or concealment to avoid attack (for example, submergence).

DECEPTION. Deliberate misrepresentation of reality to gain an advantage.

DEFENSIVE FIREPOWER. The means of destroying attacking missiles, aircraft, or torpedoes.

DEFENSIVE FORCE. The capacity to either destroy attacking weapons or defeat them by "softkill" methods other than shooting down.

ESCORTING. Actions taken by accompanying forces to protect other forces or shipping by destroying the enemy or threatening his destruction.

FIGHTING POWER. A composite of force and counterforce, representing in some way the deliverable firepower over the combat life of the fighting unit. When **firepower** and **staying power** were common terms, **fighting power** incorporated both in a quantitative representation.

FIREPOWER. The material means of a fighting unit to reduce enemy forces. It is the capacity to destroy, measured in rate of delivery (as, for instance, shells per minute, or missiles in a complete salvo).

FIREPOWER KILL. The elimination of an enemy force's means of delivering firepower for the duration of a battle.

FLEET. Major forces used to gain, maintain, or dispute control of the seas. By this definition, neither amphibious forces nor ballistic-missile submarines constitute a fleet.

FORCE. In general, the means of gaining an objective. *Military force* is the means of destroying the enemy's capacity to apply force. (Depending on the context, force is composed of ships and aircraft; guns, missiles, torpedoes, mines, and other such means of destruction; or forces plus the moral and intangible resources to destroy the enemy's capacity to apply force.)

FORCES. Units that carry force, here defined to include not only firepower but also scouting and C^2 capabilities.

MANEUVER. Movement to achieve a tactical advantage. (Maneuver may be associated with force, counterforce, scouting, or antiscouting. Ideally, maneuvers are made with all four elements in mind.)

POWER. The rate at which force may be applied against an enemy.

SCOUTING. Acts of search, detection, tracking, targeting, and enemy damage assessment, including reconnaissance, surveillance, signals intelligence, and all other means of gathering information that may be used in combat. Scouting is not accomplished until the information is delivered to the commander being served.

SCREENING. The use of forces to help protect other more valued units, accomplished by some combination of antiscouting and escorting, and often by scouting as well.

SEARCH. The sensing phase of scouting. It may be active or passive or both.

SENSING. Roughly equivalent to **scouting**, sensing makes no assumptions about whether the sensed object is recognized (the classification step) or whether the information is delivered to the commander and assimilated into the decision process.

STAYING POWER. The capacity to absorb damage and continue fighting with measurable effectiveness.

STRIKING POWER. The material means of a force to reduce enemy forces.

Appendix B

Principles of War

Two compilations of principles of war are included in this appendix. They are transcribed just as they are found. The first is in then Captain Stuart Landersman's study, *Principles of Naval Warfare,* completed in 1982 when he was a member of the special studies group in the Center for Naval Warfare Studies at the Naval War College, Newport.*

The second list is from Barton Whaley's study of deception and surprise, *Strategem,* completed in 1969.† Because he is interested in the importance given to surprise by each authority, Whaley places the principles in the order of priority the author intended or the order Whaley believes the author might have ranked them.

Landersman's Compilation of Principles of War

Sun Tzu	350 BC	Objective, unity, deception, initiative, adaptability, environment, security

*Landersman, appendix E.
†Whaley, pp. 122–26.

Napoleon	1822	Objective, offense, mass, movement, concentration, surprise, security
Clausewitz	1830	Objective, offensive, concentration, economy, mobility, surprise
Jomini	1836	Objective, maneuver, concentration, offense, deception
Mahan	1890	Objective, concentration, offense, mobility, command
Fuller	1912	Objective, mass, offense, security, surprise, movement
Foch	1918	Objective, offense, economy, freedom, disposal, security
Corbett	1918	Objective, concentration, flexibility, initiative, mobility, command
U.S. Army	1921	Objective, offense, mass, economy, movement, surprise, security, simplicity, cooperation
Nimitz	1923	Concentration, time, initiative, surprise, mobility, objective, command, environment
Fuller	1924	Objective, offense, surprise, concentration, economy, security, mobility, cooperation
Liddell Hart	1925	Objective, offense, defense, mobility
Falls	1943	Objective, concentration, protection, surprise, reconnaissance, mobility
Stalin	1945	Objective, stability, morale, divisions, armament, organization
USSR	1953	Objective, surprise, speed, coordination, attack
U.S. Navy	1955	Objective, morale, simplicity, control, offensive, exploitation, mobility, concentration, economy, surprise, security, readiness
Eccles	1965	Objective, offensive, concentration, mobility, economy, cooperation, security, surprise, simplicity
Keener	1967	Objective, distribution, coordination, initiative, surprise

Mao	1967	Objective, concentration, annihilation, mobility, offense, surprise, attack, autonomy, unity, morale
U.S. Army	1968	Objective, offense, mass, economy, maneuver, unity, security, surprise, simplicity
Royal Navy	1969	Aim, morale, offense, security, surprise, concentration, economy, flexibility, cooperation, administration
Gorshkov	1976	Scope, strike, battle, interaction, maneuver, speed, time, dominance
Hayward	1976	Scope, strike, technology, mobility, coordination, readiness, concentration, reserve

Whaley's Compilation of Principles of War, c. B.C. to A.D. 1968

	Order of Priority								
Theoretician	*1*	*2*	*3*	*4*	*5*	*6*	*7*	*8*	*9*
Sun Tzu 4th cent. B.C.	Objective	Offensive	Surprise	Concentration	Mobility	Coordination			
Vegetius ca. 390 A.D.	Mobility	Security	Surprise	Offensive					
Saxe 1757	Mobility	Morale	Security	Surprise					
Napoleon 1822	Objective	Offensive	Mass	Movement	Surprise	Security			
Clausewitz 1832	Objective	Offensive	Concentration	Economy of Force	Mobility	Surprise			
Jomini 1836	Objective	Movement	Concentration	Offensive	Diversion				
P. L. MacDougall 1858	Mass	Direction							
N. B. Forrest 1864	Mass	Direction	Rapidity	Offensive					
Fuller 1912	Objective	Mass	Offensive	Security	Surprise	Movement			
Stalin (1918–47)	Stability of the rear	Morale	Quality and quantity	Armament	Organizing ability of commanders	Surprise			

Whaley's Principles (*continued*)

Theoretician				*Order of Priority*					
	1	*2*	*3*	*4*	*5*	*6*	*7*	*8*	*9*
Foch 1918	Offensive	Economy of force	Freedom of action	Free disposal of forces	Security				
C. V. F. Townshend 1920	Objective	Economy of force	Mass	Offensive	Direction	Security			
U.S. War Dept., *Training Regulations*, nos. 10–5 1921	Objective	Offensive	Mass	Economy of force	Movement	Surprise	Security	Simplicity	Cooperation
Fuller 1925	Direction	Offensive	Surprise	Concentration	Distribution	Security	Mobility	Endurance	Determination
Liddell Hart 1929	Objective	Movement	Surprise						
U.S. Command and General Staff School 1936	Offensive	Concentration	Economy of force	Mobility	Surprise	Security	Cooperation		
Mao 1938	Political objective	Mobility	Offensive	Defensive	Concentration	Surprise			

Whaley's Principles (*continued*)

	Order of Priority								
Theoretician	*1*	*2*	*3*	*4*	*5*	*6*	*7*	*8*	*9*
U.S. Army FM 100-5 1941, 1944	Objective	Simplicity	Unity of com-mand	Offensive	Con-centra-tion of superior force	Surprise	Security		
Cyril Falls ca. 1945	Economy of force	Protection	Surprise	Aggres-sive re-connais-sance	Mainte-nance of the aim				
Liddell Hart (1954–67)	Alterna-tive ob-jectives	Move-ment	Surprise						
Giap 1960	Political Objec-tive	Speed	Surprise	Morale	Security	Cooper-ation			
Guevara 1960	Objective	Mobility	Surprise						
Montgomery 1968	Surprise	Concen-tration of effort	Coopera-tion of all arms	Control	Simplicity	Speed of action	Initiative		
U.S. Army FM 100-5 (1962–68)	Objective	Offensive	Mass	Economy of force	Maneuver	Unity of com-mand	Security	Surprise	Simplicity

Since both Landersman and Whaley have studied the principles of war far more extensively than I, their lists must speak for themselves. But I find the differences, both within and between the two lists, as instructive as the similarities and therefore worthy of comment. It is fascinating to contemplate the reasons for these differences:

—The age and maturity of the author. Individuals' lists as well as organizations' lists may change over time.

—The historical period. The size of states, their forces, and their weapons affect the selection and priority of principles.

—The social milieu, especially whether it is Oriental or Occidental.

—The emphasis on tactics, strategy, or both. Nimitz deals explicitly with tactics, for example.

—The author's experience or viewpoint. Was it centered on international or revolutionary war? On major battles or guerrilla war?

—The military milieu, for example, whether the conflict takes place on land or at sea. (It seems that no author has emphasized air or amphibious warfare.)

As Whaley says, citing Henry Eccles,* these lists imply empirical maxims for a commander to bear in mind rather than principles denoting a primary source of wisdom, a fundamental truth, or a comprehensive law or doctrine from which others are derived. They are lists of key words which in conjunction contain rich meaning. In many cases, they are interpretations of the author's intent, made either by Landersman or Whaley or by some intermediate student of the author.

Take Nimitz, for example. As a middle-grade officer with experience in World War I behind him, he derived the principles recorded below from a slightly abbreviated transcript of his Naval War College thesis, completed in the spring of 1923. While Landersman has reduced Nimitz's conclusions to a set of representative key words, or "principles," the transcript itself is much more comprehensive, and even that is only a summary. Key words are a useful summation of a wise man's deductions about the *principles* of war. Simple mathematical equations and graphical depictions are a useful summation

*Eccles, pp. 108–13.

of the *process* of war. I hasten to add that my friend Stu Landersman made much the same point in his study.

Thesis on Tactics*

The main and unchanging principles of warfare are:

FIRST: To employ *all* the forces which can be made available with the utmost energy. (This does not necessarily imply the offensive with its attendant advantages.)

SECOND: To concentrate superior forces against the enemy at the point of contact or where the decisive blow is to be struck.

THIRD: To avoid loss of time.

FOURTH: To follow up every advantage gained with utmost energy.

MAIN PRINCIPLES:

(a) Attempt to surprise and deceive the enemy as to the plan of battle, and method and point of attack.

(b) Endeavor to isolate a portion of the enemy battle line and crush it before it can be supported.

(c) Maneuver on interior lines to save time, increase mobility and facilitate concentration.

(d) Plan the battle so as to cut off retreat in case your force is stronger, or to facilitate breaking off the action if your force is the weaker.

(e) Adhere to the plan and do not lose sight of the objective.

(f) Modern fleets cannot be handled in single line by one officer. They must be in subdivisions, all within supporting distance of each other, each subdivision controlled by a subordinate upon whom must be imposed authority, responsibility, and great freedom of initiative in accomplishing the end in view.

(g) Make all practicable use of such natural advantages as may be obtained from the direction of the wind, the state of the sea, direction of the sun, fogs and reduced visibility, smoke, and smoke screens.

(h) As a general rule, great results cannot be accomplished without a corresponding degree of risk. Efficient fleets are never *perfectly* ready for action. The leader who awaits perfection of plans, material, or training, will wait in vain, and in the end will yield the victory to him who employs the tools at hand with the greatest vigor.

*Nimitz, pp. 3–4.

Bibliography

Compared with the number of books on strategy, the unclassified books in English on naval tactics are very few. The same may be said of articles in periodicals and other published research. But if one looks for comments, either explicit or implicit, in studies and histories of war, in anthologies of sea stories, in specialized books on, say, the development of aircraft carriers, radar, or naval architecture, or in books on naval operations analysis, the potential number of tactical insights is large indeed.

This bibliography is as complete as possible in one and only one respect: it includes all studies of naval tactics in English known to me, whatever their merit. In all other respects it is highly selective, listing books, articles, and papers for one of two reasons: either the work was cited in the text or it was thought especially worthy of consultation by anyone undertaking research into the history, techniques, processes, combat environment, or analysis of tactics.

Abchuck, V. G., et al. *Vyendenue v Teoriu Vyraborki Reshenii* (Introduction to decision-making theory). Moscow: Voyenizdat, 1972.

Albion, Robert G. *Makers of Naval Policy, 1798–1947*. Annapolis, Maryland: Naval Institute Press, 1980.

Allen, Captain Charles D. (USN Ret.). "Forecasting Future Forces." U.S. Naval Institute *Proceedings* (Nov 1982).

Bainbridge-Hoff, Commander William (USN). *Elementary Naval Tactics*. New York: John Wiley, 1894.

———. *Examples, Conclusions, and Maxims of Modern Naval Tactics*. Washington, D.C.: U.S. Government Printing Office, 1884.

Baudry, Lieutenant Ambroise (French navy). *The Naval Battle: Studies of Tactical Factors*. London: Hughes Rees, Ltd., 1914.

Beesly, Patrick. *Very Special Intelligence: The Story of the Admiralty's Operational Intelligence Centre, 1939–1945*. London: Hamish Hamilton, 1977; New York: Ballantine Books, 1981.

Belot, Admiral Raymond de (French navy). *The Struggle for the Mediterranean, 1939–1945*. Translated by J. A. Field. Princeton: Princeton University Press, 1951.

Bernotti, Lieutenant Romeo (Italian navy). *The Fundamentals of Naval Tactics*. Annapolis, Maryland: U.S. Naval Institute, 1912.

Blackett, P. M. S. *Studies of War*. New York: Hill and Wang, 1962.

Brodie, Bernard. *A Layman's Guide to Naval Strategy*. Princeton: Princeton University Press, 1942.

———. *Sea Power in the Machine Age*. Princeton: Princeton University Press, 1943.

———, and Fawn Brodie. *From Crossbow to H-Bomb*. Rev. ed. Bloomington, Indiana: Indiana University Press, 1973.

Bush, Vannever. *Modern Arms and Free Men*. New York: Simon & Schuster, 1949.

Clausewitz, Carl von. *On War*. Edited and translated by Michael Howard and Peter Paret. Princeton: Princeton University Press, 1976.

Clowes, Sir William Laird. *The Royal Navy: A History*. 7 vols. London: Sampson Low, 1897–1903.

Corbett, Sir Julian S. *Some Principles of Maritime Strategy*. London: Longmans, Green & Co., 1911.

Creswell, John. *British Admirals of the Eighteenth Century: Tactics in Battle*. Hamden, Connecticut: Archon Books, 1972.

Creveld, Martin van. *Command in War*. Cambridge, Massachusetts: Harvard University Press, 1985.

Cushman, Lieutenant General John H. (USA Ret.). *Command and Control of Theater Forces: Adequacy*. Cambridge, Massachusetts: Harvard University Press, 1983.

d'Albos, Emmanuel E. A. *Death of a Navy: Japanese Naval Action in World War II*. New York: Devin-Adair, 1957.

Daniel, Donald C. *Antisubmarine Warfare and Superpower Strategic Stability*. London: The Macmillan Co., 1985.

———, and Katherine L. Herbig, eds. *Strategic Military Deception*. New York: Pergamon Press, 1982.

Davis, Vincent. "The Politics of Innovation: Patterns in Navy Cases." Vol. 4, monograph no. 3 in *Monograph Series in World Affairs*. Denver, Colorado: University of Denver, 1967.

Deitchman, Seymour J. *New Technology and Military Power: General Purpose Forces for the 1980s and Beyond*. Boulder, Colorado: Westview Press, 1979.

Deuterman, Commander P. T. (USN). "The Matched Pair: A Tactical Concept." U.S. Naval Institute *Proceedings* (Jan 1982).

Douglas, Joseph D., Jr., and Amoretta A. Hoeber. "The Role of the U.S. Surface Navy in Nuclear War." U.S. Naval Institute *Proceedings* (Jan 1982).

Dull, Paul S. *A Battle History of the Imperial Japanese Navy, 1941–1945*. Annapolis, Maryland: Naval Institute Press, 1978.

Dunnigan, James F. *How to Make War: A Comprehensive Guide to Modern War*. New York: Morrow, 1982.

Dupuy, Colonel Trevor N. (USA Ret.) *Understanding War: History and Theory of Combat.* New York: Paragon House, 1987.

———. *Numbers, Predictions, and War: Using History to Evaluate Combat Factors and Predict the Outcomes of Battles*. Indianapolis, Indiana: Bobbs-Merrill, 1979.

Eccles, Henry C. *Military Concepts and Philosophy*. New Brunswick, New Jersey: Rutgers University Press, 1965.

Fioravanzo, Admiral Giuseppe (Italian navy). *A History of Naval Tactical Thought*. Translated by Arthur W. Holst. Annapolis, Maryland: Naval Institute Press, 1979.

Fiske, Commander Bradley A. (USN). "American Naval Policy." U.S. Naval Institute *Proceedings* (Jan 1905).

Friedman, Norman. *Naval Radar*. Greenwich, England: Conway Maritime Press, 1981.

———. *U.S. Aircraft Carriers: An Illustrated History*. Annapolis, Maryland: Naval Institute Press, 1983.

———. *U.S. Destroyers: An Illustrated Design History*. Annapolis, Maryland: Naval Institute Press, 1982.

Frost, Holloway Halstead. *The Battle of Jutland*. Annapolis, Maryland: U.S. Naval Institute, 1936.

Fuller, J. F. C. *The Conduct of War, 1789–1961*. New Brunswick, New Jersey: Rutgers University Press, 1961.

Genda, General [*sic*] Minoru, 3 JSDF (Ret.). "Tactical Planning in the Imperial Japanese Navy." Naval War College *Review* (Oct 1962).

Gooch, John, and Amos Perlmutter. *Military Deception and Strategic Surprise*. Totowa, New Jersey: Frank Cass & Co, Ltd., 1982.

Gorshkov, Admiral of the Fleet S. G. (Soviet navy). *The Development of the Art of Naval Warfare*. Translated by T. A. Neely, Jr. U.S. Naval

Institute *Proceedings* (June 1975). First printed in *Morskoy Sbornik* (no. 12, 1974).

Grenfell, Captain Russell (RN). *Nelson the Sailor*. New York: The Macmillan Co., 1950.

Gretton, Sir Peter (RN). *Crisis Convoy: The Story of HX231*. Annapolis, Maryland: Naval Institute Press, 1974.

Hackett, General Sir John, et al. *The Third World War: A Future History*. New York: The Macmillian Co., 1978.

Hazen, David C. "Nine Prejudices About Future Naval Systems." U.S. Naval Institute *Proceedings* (July 1980).

Hough, Richard A. *Dreadnought*. New York: The Macmillan Co., 1964.

———. *The Great War at Sea, 1914–18*. Oxford: Oxford University Press, 1983.

Hughes, Terry, and John Costello. *The Battle of the Atlantic*. New York: The Dial Press/James Wade, 1977.

Hughes, Wayne P., Jr. "Speed Characteristics of the Treaty Cruisers." U.S. Naval Institure *Proceedings* (Feb 1953).

Hwang, John, Daniel Schuster, Kenneth Shere, and Peter Vena, eds. *Selected Analytical Concepts in Command and Control*. New York: Gordon & Breach, 1982.

Ivanov, D. A., V. P. Savel'yev, and P. V. Shemanskiy. *Osnovy Upravleniya Voyskami v Boyu* (Fundamentals of troop control at the tactical level). Moscow: Voyenizdat. Translation published by U.S. Government Printing Office, Washington, D.C., 1983.

Jameson, Rear Admiral William (RN). *The Fleet That Jack Built: Nine Men Who Made a Modern Navy*. London: Rupert Hart-Davis, 1962.

Kahn, David. *The Code Breakers*. New York: The Macmillan Co., 1967.

Kelsey, Commander Robert J. (USN). "Maneuver Warfare at Sea." U.S. Naval Institute *Proceedings* (Sept 1982).

Kemp, Peter, ed. *The Oxford Companion to Ships and the Sea*. London: Oxford University Press, 1976.

Kennedy, Paul M. *The Rise and Fall of British Naval Mastery*. New York: Charles Scribner's Sons, 1976.

Koopman, Bernard O. *Search and Screening: General Principles with Historical Applications*. Elmsford, New York: Pergamon Press, 1980.

Lanchester, Frederick W. "Mathematics in Warfare." In *The World of Mathematics*, edited by James R. Newman. New York: Simon and Schuster, 1956.

Landersman, Captain Stuart (USN). *Principles of Naval Warfare*. Newport, Rhode Island: Naval War College, 1982.

Lanza, Conrad H. *Napoleon and Modern War: His Military Maxims*. Harrisburg, Pennsylvania: Military Service Publishing Co., 1943.

Lautenschläger, Karl. "Technology and the Evolution of Naval Warfare,

1851–2001." Charles H. Davis Series Spring Lecture, U.S. Naval Postgraduate School, Monterey, California, April 1984. Washington, D.C.: National Academy Press, 1984.

Lehman, John. *Aircraft Carriers: The Real Choices*. Beverly Hills, California: Sage Publications, 1978.

Levert, Lee J. *Fundamentals of Naval Warfare*. New York: The Macmillan Co., 1947.

Lewin, Ronald. *The American Magic: Codes, Ciphers, and the Defeat of Japan*. Great Britain: Hutchinson & Co. Ltd., 1982; New York: Penguin Books, 1983.

———. *Ultra Goes to War*. New York: McGraw-Hill Book Co., 1978.

Lewis, Michael. *The History of the British Navy*. Baltimore, Maryland: Pelican Books, 1957.

———. *The Navy of Britain: A Historical Portrait*. London: George Allen and Unwin, 1948.

Liddell Hart, B. H. *Strategy*. London: Faber and Faber, 1967; New York: Signet, 1974.

McHugh, Francis J. *Fundamentals of War Gaming*, 3rd ed. Newport, Rhode Island: Naval War College, 1966.

MacIntyre, Donald, and Basil W. Bathe. *Man-of-War: A History of the Combat Vessel*. New York: McGraw-Hill Book Co., 1969.

McKearney, Lieutenant Commander Terrance J. (USN). "The Solomons Naval Campaign: A Paradigm for Surface Warships in Maritime Strategy."

Mahan, Alfred Thayer. *The Influence of Sea Power on History, 1660–1783*. Boston: Little Brown, 1890.

Makaroff, Vice Admiral S. O. (Russian navy). *Discussion of Questions in Naval Tactics*. Translated by Lieutenant John B. Bernadou (USN). ONI, part 2, General Information Series, no. 17. Washington, D.C.: Government Printing Office, 1898.

Marble, Ensign Frank (USN). "The Battle of the Yalu." U.S. Naval Institute *Proceedings* (fall 1895).

Melhorn, Charles M. *Two-Block Fox: The Rise of the Aircraft Carrier, 1911–1929*. Annapolis, Maryland: Naval Institute Press, 1974.

Mitchell, Donald W. *History of the Modern American Navy from 1883 Through Pearl Harbor*. New York: A. Knopf, 1946.

Mordal, Jacques. *Twenty-Five Centuries of Sea Warfare*. Translated by Len Ortzen. London: Souvenir Press, 1965.

Morison, Elting E. *Admiral Sims and the Modern American Navy*. Boston: Houghton-Mifflin, 1942.

———. *Men, Machines, and Modern Times*. Cambridge, Massachusetts: The MIT Press, 1966.

———. "The Navy and the Scientific Endeavor." *Science and the Future Navy: A Symposium*. Washington, D.C.: National Academy of Sciences, 1977.

Morison, Samuel Eliot. *History of United States Naval Operations in World War II*. 15 vols. Boston: Little, Brown, and Company, 1947–62.

Musashi, Miyamoto. *A Book of Five Rings*. Translated by Victor Harris. Woodstock, New York: Overlook Press, 1974.

Nimitz, Chester W. "Thesis on Tactics." Newport, Rhode Island: Naval War College, 1923.

Osipov, M. "Vlyeyanye Chislyennosti Srazhayush-chiksya Storen Na Ix Potyera" (The influence of the numerical strength of engaged sides on their losses). *Voenniy Sbornik* (Military collection): (no. 6, June 1915; no. 7, July 1915; no. 8, Aug 1915; no. 9, Sep 1915; no. 10, Oct 1915).

Palmer, Joseph. *Jane's Dictionary of Naval Terms*. London: Macdonald & Janes, 1975.

Paxson, E. W., M. G. Weiner, and R A. Wise. "Interactions Between Tactics and Technology in Ground Warfare." Rand Report R-2377-ARPA, Santa Monica, California, Jan 1979.

Pemsel, Helmut. *A History of War at Sea: An Atlas and Chronology of Conflict at Sea from Earliest Times to the Present*. Translated by D. G. Smith. Annapolis, Maryland: Naval Institute Press, 1975.

Polmar, Norman. *Aircraft Carriers: A Graphic History of Carrier Aviation and Its Influence on World Events*. Garden City, New York: Doubleday & Co., 1969.

Potter, Elmer B., ed. *Sea Power: A Naval History*. 2d edition. Annapolis, Maryland: Naval Institute Press, 1981.

Pratt, Fletcher. *Night Work: The Story of Task Force 39*. New York: Henry Holt, 1946.

———. *Our Navy: A History*. Garden City, New York: Garden City Publishing Co., 1941.

Raven, Alan, and John Roberts. *British Battleships of World War II*. Annapolis, Maryland: Naval Institute Press, 1976.

Reynolds, Clark G. *Command of the Sea: The History and Strategy of Maritime Empires*. New York: William Morrow, 1974.

———. *The Fast Carriers: The Forging of An Air Navy*. New York: McGraw-Hill, 1968.

Richten, Eberhardt. "The Technology of Command." *Naval War College Review* (March–April 1984).

Robison, Rear Admiral Samuel S. (USN), and Mary L. Robison. *A History of Naval Tactics from 1530 to 1930*. Annapolis, Maryland: U.S. Naval Institute, 1942.

Rohwer, Jurgen. *The Critical Convoy Battles of March 1943*. Annapolis, Maryland: Naval Institute Press, 1977.

Roskill, Captain Stephen W. (RN). *The War at Sea, 1939–1945*. 3 vols. London: H. M. Stationery Office, 1954–56.

———. *White Ensign: The British Navy at War, 1939–1945*. Annapolis, Maryland: U.S. Naval Institute, 1960.

Sanderson, Michael. *Sea Battles: A Reference Guide*. Middletown, Connecticut: Wesleyan University Press, 1975.

Seaquist, Commander Larry (USN). "Tactics to Improve Tactical Proficiency." U.S. Naval Institute *Proceedings* (Feb 1983).

Secretary of State for Defense. *The Falklands Campaign: The Lessons*. London: H.M. Stationery Office, 1982.

Secretary of the Navy's Task Force Report. *South Atlantic Conflict Lessons Learned*. Washington, D.C.: Navy Department, 1983.

Spector, Ronald H. *Eagle Against the Sun: The American War with Japan*. New York: The Free Press, The Macmillan Co., 1985.

Stafford, Commander Edward P. (USN). *The Big E: The Story of the USS Enterprise*. New York: Random House, 1962.

Stalbo, Vice Admiral K. "Some Issues of the Theory of the Development and Deployment of the Navy." *Morskoy Sbornik*. Nos. 4 and 5, 1981.

Sternhell, Charles M., and Alan M Thorndike. *OEG Report No 51: Antisubmarine Warfare in World War II*. Washington, D.C.: OEG, Office of the CNO, Navy Department, 1946.

Talbott, J. E. "Weapon Development, War Planning and Policy: The U.S. Navy and the Submarine, 1917–41. *Naval War College Review* (May–June 1984).

Tanaka, Rear Admiral Raizo (Japanese navy). "The Struggle for Guadalcanal." In *The Japanese Navy In World War II*, edited by Raymond O'Connor. Annapolis, Maryland: Naval Institute Press, 1969.

Taylor, James G. *Initial Concept of Soviet C^2*. Monterey, California: U.S. Naval Postgraduate School, 1984.

Taylor, Theodore C. "A Basis for Tactical Thought." U.S. Naval Institute *Proceedings* (June 1982).

Tidman, Keith R. *The Operations Evaluation Group: A History of Naval Operations Analysis*. Annapolis, Maryland: Naval Institute Press, 1984.

Tzu, Sun. *The Art of War*. Translated by Samuel B. Griffith. London: Oxford University Press, 1963.

Uhlig, Frank, Jr. "Naval Tactics: Examples and Analogies." *Naval War College Review* (March–April, 1981).

United States Air Force. *Dictionary of Basic Military Terms: A Soviet View*. Washington, D.C.: U.S. Government Printing Office, 1976.

United States Joint Chiefs of Staff. *Department of Defense Dictionary of Military and Associated Terms*. JCS publication 1. Washington, D.C.: U.S. Government Printing Office, 1984.

Vlahos, Michael, *Blue Sword*. Newport, Rhode Island: Naval War College, 1980.

Washburn, Alan R. "Gross Measures of Surface-to-Surface Naval Firepower." Monterey, California: U.S. Naval Postgraduate School, 1978.

Watson, Bruce W., and Peter M. Dunn, eds. *Military Lessons of the Falkland Islands War*. Boulder, Colorado: Westview Press, 1984.

Whaley, Barton. *Strategem: Deception and Surprise in War*. Cambridge, Massachusetts: MIT Center for International Studies, 1969.

Wiener, Norbert. *Cybernetics*. 2d ed. Cambridge, Massachusetts: The MIT Press, 1961.

Willmott, H. P. *The Barrier and the Javelin: Japanese and Allied Pacific Strategies, February to June 1942*. Annapolis, Maryland: Naval Institute Press, 1983.

Wilson, Henry W. *Ironclads in Action*. 2 vols. Boston: Little Brown, 1896.

Woodward, David. *The Russians at Sea*. London: Kimber, 1965.

Index

Note: Terms in boldface are defined or discussed in appendix A; page numbers in italics indicate terms defined in the text.

The **Naval Institute Press** is the book-publishing arm of the U.S. Naval Institute, a private, nonprofit professional society for members of the sea services and civilians who share an interest in naval and maritime affairs. Established in 1873 at the U.S. Naval Academy in Annapolis, Maryland, where its offices remain today, the Naval Institute has more than 100,000 members worldwide.

Members of the Naval Institute receive the influential monthly magazine *Proceedings* and discounts on fine nautical prints, ship and aircraft photos, and subscriptions to the quarterly *Naval History* magazine. They also have access to the transcripts of the Institute's Oral History Program and get discounted admission to any of the Institute-sponsored seminars regularly offered around the country.

The Naval Institute's book-publishing program, begun in 1898 with basic guides to naval practices, has broadened its scope in recent years to include books of more general interest. Now the Naval Institute Press publishes more than forty new titles each year, ranging from how-to books on boating and navigation to battle histories, biographies, ship and aircraft guides, and novels. Institute members receive discounts on the Press's more than 375 books.

Full-time students are eligible for special half-price membership rates. Life memberships are also available.

For a free catalog describing the Naval Institute Press books currently available, and for further information about U.S. Naval Institute membership, please write to:

The **Naval Institute Press** is the book-publishing arm of the U.S. Naval Institute, a private, nonprofit professional society for members of the sea services and civilians who share an interest in naval and maritime affairs. Established in 1873 at the U.S. Naval Academy in Annapolis, Maryland, where its offices remain today, the Naval Institute has more than 100,000 members worldwide.

Members of the Naval Institute receive the influential monthly magazine *Proceedings* and discounts on fine nautical prints, ship and aircraft photos, and subscriptions to the quarterly *Naval History* magazine. They also have access to the transcripts of the Institute's Oral History Program and get discounted admission to any of the Institute-sponsored seminars regularly offered around the country.

The Naval Institute's book-publishing program, begun in 1898 with basic guides to naval practices, has broadened its scope in recent years to include books of more general interest. Now the Naval Institute Press publishes more than forty new titles each year, ranging from how-to books on boating and navigation to battle histories, biographies, ship and aircraft guides, and novels. Institute members receive discounts on the Press's more than 375 books.

Full-time students are eligible for special half-price membership rates. Life memberships are also available.

For a free catalog describing the Naval Institute Press books currently available, and for further information about U.S. Naval Institute membership, please write to: